Specification-Driven Product Development

Specification-Driven Product Development

Edward K. Bower

iUniverse, Inc.
New York Lincoln Shanghai

Specification-Driven Product Development

All Rights Reserved © 2003 by Edward K. Bower

No part of this book may be reproduced or transmitted in any form or by any means, graphic, electronic, or mechanical, including photocopying, recording, taping, or by any information storage retrieval system, without the written permission of the publisher.

iUniverse, Inc.

For information address:
iUniverse, Inc.
2021 Pine Lake Road, Suite 100
Lincoln, NE 68512
www.iuniverse.com

ISBN: 0-595-27185-5

Printed in the United States of America

Contents

CHAPTER 1 INTRODUCTION 1
- *1.1 Product Development Methods 1*
- *1.2 Definitions of Terms 3*
- *1.3 Assumed Development Environment 4*
- *1.4 Overview 5*

CHAPTER 2 INFORMAL DEVELOPMENT 7

CHAPTER 3 FORMAL DEVELOPMENT 11
- *3.1 A Historic Parallel 12*
- *3.2 Development Phases 12*
- *3.3 Sequence of Phases 13*
- *3.4 Formal Documentation 13*
- *3.5 Functional Changes 14*
- *3.6 Formal Methods Are Counterintuitive 16*
- *3.7 Bower's Inequality 17*
- *3.8 Bower's Equation 18*

CHAPTER 4 THE FUNCTIONAL SPECIFICATION 19
- *4.1 Storage of Product Functional Information 19*
- *4.2 Access to Functional Information 20*
- *4.3 Information Access Problems 20*
  - 4.3.1 Constant research is needed by many people 21
  - 4.3.2 Repeated research may yield different answers 22
  - 4.3.3 Decisions are delayed 23
  - 4.3.4 Resources are consumed 23

- 4.4 The Common-Sense Solution 24
- 4.5 The Fundamental Rule 25
- 4.6 Generation of the Functional Specification 25
  - 4.6.1 Marketing contributions 27
  - 4.6.2 Roles of other groups 29
  - 4.6.3 Sources of functional information 30
  - 4.6.4 FS authors 31
  - 4.6.5 FS reviewers 31
- 4.7 Approval of the FS 32
- 4.8 Changes to the FS 33
  - 4.8.1 Why are changes bad? 33
  - 4.8.2 Why are changes necessary? 36
  - 4.8.3 Change control 37
  - 4.8.4 The Modification Request (MR) 37
  - 4.8.5 Review of MRs 39
  - 4.8.6 FS modification 39
  - 4.8.7 Re-approval of the modified FS 40
- 4.9 Violations of the Fundamental Rule 40
  - 4.9.1 Sneak paths 41
  - 4.9.2 Managerial functional decisions 43
  - 4.9.3 Parallel FS and Wish List 44
  - 4.9.4 Collected FS changes 47
  - 4.9.5 Evolutionary development 50
  - 4.9.6 Inter-departmental leaks 52
- 4.10 Other Problem Areas 54
  - 4.10.1 Winning employee cooperation 54
  - 4.10.2 Continual refinements 58
  - 4.10.3 If agreement can't be reached 58
  - 4.10.4 The role of prototyping 59
- 4.11 Other Business Models 62

CHAPTER 5   WHAT'S IN A FUNCTIONAL
            SPECIFICATION? 64

- *5.1 Boundary Definition* *65*
- *5.2 Complete Functional Requirements* *66*
- *5.3 Formal Language* *67*
- *5.4 Reference to Available Standards* *68*
- *5.5 Consistent Terminology* *70*
- *5.6 Logical Organization* *70*
- *5.7 Numbered Sections* *71*
- *5.8 The Product's Name* *71*

CHAPTER 6   WHAT'S NOT IN A FUNCTIONAL SPECIFICATION? 73

- *6.1 Anything that Isn't a Function of the Product* *73*
  - 6.1.1 HOW the product works 73
  - 6.1.2 Wishes 74
  - 6.1.3 Extraneous topics 75
  - 6.1.4 Specification of the design process 75
- *6.2 Duplication* *76*
- *6.3 Bullets* *78*
- *6.4 Selective Emphasis* *78*
- *6.5 Untestable Requirements* *79*
- *6.6 Contradictions* *79*
- *6.7 Impossible Precision* *79*
- *6.8 Unnecessary Requirements* *80*
- *6.9 Over-specification* *80*
- *6.10 Proprietary Information* *82*
- *6.11 Open Issues* *82*

CHAPTER 7   OTHER DEVELOPMENT DOCUMENTS 83

- *7.1 Product Development Procedure* *83*
  - 7.1.1 Applicability, authorization, responsibility, and flexibility 84
  - 7.1.2 Define a formal method 87
  - 7.1.3 Definition of phases 87

      7.1.4 Definition of documents 88
      7.1.5 Description of reviews 90
      7.1.6 The Document Control System 90
      7.1.7 The Engineering Process Group 91
- *7.2 Wish List* 91
- *7.3 Architecture Specification* 91
- *7.4 High-Level Design Specification* 95
- *7.5 Detailed Design Specification* 96
- *7.6 Development Plan* 97
- *7.7 Acceptance Test Procedure and Test Report* 97
- *7.8 Modification Request* 98
- *7.9 Future Feature List.* 99
- *7.10 Meeting Minutes* 99

CHAPTER 8   THE REVIEW PROCESS 100
- *8.1 Preparation for a Review* 100
- *8.2 The Purpose of a Review* 101
- *8.3 Conducting a Review Meeting.* 101
- *8.4 After the Review.* 103
- *8.5 Specific Techniques to Review an FS* 104
      8.5.1 Reviewer as User of the product 105
      8.5.2 Reviewer as Customer for the product 105
      8.5.3 Reviewer as Buyer of the product 106
      8.5.4 Reviewer as Seller of the product 108
      8.5.5 Reviewer as Buyer of contracted development 108
      8.5.6 Reviewer as Development Contractor. 110
      8.5.7 Reviewer as Middleman 111
- *8.6 Specific Techniques to Review Design Specifications* 111

CHAPTER 9   DEVELOPMENT PHASES 114
- *9.1 Definition Phase* 115
- *9.2 Architecture/Planning Phase* 115
- *9.3 High-Level Design Phase* 115

Contents ix

- *9.4 Detailed Design Phase* *116*
- *9.5 Implementation Phase* *116*
- *9.6 Acceptance Test Phase* *117*
- *9.7 Summary* *118*

CHAPTER 10 THE QUALITY PROGRAM AND CERTIFICATION 120

- *10.1 Basic Quality Principle* *120*
- *10.2 Verification and Validation* *121*
- *10.3 Reasons for Certification* *123*
- *10.4 A Caution* *124*

CHAPTER 11 SOME ESTABLISHED FORMAL METHODS 126

- *11.1 ISO 9001* *126*
- *11.2 Capability Maturity Model* *128*
  - 11.2.1 Level 1 — Initial 128
  - 11.2.2 Level 2 — Repeatable 129
  - 11.2.3 Level 3 — Defined 130
  - 11.2.4 Level 4 — Managed 132
  - 11.2.5 Level 5 — Optimizing 132
- *11.3 NASA Standards* *133*
- *11.4 IEEE Standards* *134*
- *11.5 Government Standards* *134*
- *11.6 Observations* *135*
  - 11.6.1 Endorsement 135
  - 11.6.2 Similarities among methods 135
  - 11.6.3 General applicability 136
  - 11.6.4 Incremental migration toward formality ... 137
  - 11.6.5 Definition of a formal method 138

CHAPTER 12 BENEFITS OF THE FORMAL PROCESS 140

- *12.1 Company-wide Benefits* *142*

- 12.1.1 The FS communicates a uniform definition to all areas 142
- 12.1.2 Documents store knowledge over time. 143
- 12.1.3 Documents are easily transported. 143
- 12.1.4 The FS makes specification changes visible. 144
- 12.1.5 The FS motivates work at the appropriate phase. 147
- 12.1.6 The FS minimizes changes during design 147
- 12.1.7 Formal methods contribute to higher product quality 148
- *12.2 Human Resources-related Benefits 148*
  - 12.2.1 The FS as a training textbook. 149
  - 12.2.2 Rapid learning by new or transferred employees. 149
  - 12.2.3 The FS as a recruiting tool 149
  - 12.2.4 The FS helps retain employees 150
- *12.3 Management and Finance-related Benefits. 151*
  - 12.3.1 Communication to Top Management and Directors 152
  - 12.3.2 Communication to potential investors 152
  - 12.3.3 Value to potential investors 153
  - 12.3.4 Sale or transfer of a product being developed 153
  - 12.3.5 High return on investment. 154
- *12.4 Manufacturing Benefits 155*
  - 12.4.1 The FS enables a quality program. 155
  - 12.4.2 The formal documents provide feature traceability. 155
  - 12.4.3 Potential for quality certification 156
  - 12.4.4 The FS enables a robust, predictable Acceptance Test Plan. 156
  - 12.4.5 Other Manufacturing benefits 157
- *12.5 Marketing and Sales-related Benefits 157*
  - 12.5.1 Formal methods can reduce Time To Market 158
  - 12.5.2 The FS separates WHAT from HOW 158
  - 12.5.3 The FS is a reference for salespersons 159
  - 12.5.4 The FS allows specific change proposals. 160
  - 12.5.5 The FS can be shown to selected customers 160
  - 12.5.6 The FS can be submitted with a bid. 161
  - 12.5.7 The FS can be compared with a customer's specification 161

- 12.5.8 The FS can be shown to distributors or sales representatives 162
- 12.5.9 The FS is the basis of the User's Manual 162
- 12.5.10 The FS is the basis of sales brochures and specification sheets 163
- 12.5.11 The FS is the basis of advertising copy...................... 163
- 12.5.12 The FS communicates to a marketing consultant or business planner... 163
- *12.6 Customer Service-related Benefits 163*
- *12.7 Engineering-related Benefits 164*
  - 12.7.1 Designers know what to design 164
  - 12.7.2 Designers know when to stop 165
  - 12.7.3 The FS is the basis for design documents..................... 165
  - 12.7.4 The formal documents are the references for design reviews........ 165
  - 12.7.5 Formal methods produce productivity gains 166
  - 12.7.6 Formal methods increase early defect detection 167
  - 12.7.7 The Development Plan is the basis for management tracking....... 167
  - 12.7.8 Formal methods reduce schedule overruns.................... 167
  - 12.7.9 Formal methods reduce development risk 168
  - 12.7.10 The FS promotes specialization of labor..................... 169
  - 12.7.11 The FS is used to solicit contract development................. 170
  - 12.7.12 The FS is the basis for a contract 170
  - 12.7.13 Suspension and resumption of development 171
  - 12.7.14 Formal documents assist maintenance 171
- *12.8 Other Benefits 172*

CHAPTER 13 DECIDING TO UPGRADE TO FORMAL METHODS 173

- *13.1 Short-run Costs vs. Long-term Benefits 173*
- *13.2 Stated Objections To Upgrading............................. 174*
  - 13.2.1 Formality takes too long 174
  - 13.2.2 Our situation is special................................... 176
  - 13.2.3 Formality stifles creativity 177
  - 13.2.4 Formality increases bureaucracy............................ 178
  - 13.2.5 Formality costs more 178

- 13.2.6 What we are using is good enough 179
- 13.2.7 Small changes require too much effort 180
- 13.2.8 Our design tool produces adequate documentation 181
- *13.3 Actual Obstacles To Upgrading.......... 181*
  - 13.3.1 Lack of understanding of formal methods 182
  - 13.3.2 Disbelief of claimed results.......... 185
  - 13.3.3 Difficulty in getting cooperation 187
  - 13.3.4 Fear of loss of control.......... 188
  - 13.3.5 Fear of promotion 192
  - 13.3.6 Fear of commitment and accountability.......... 193
  - 13.3.7 Fear of failure.......... 195
  - 13.3.8 Fear of success 195
  - 13.3.9 Admission of past misrepresentation.......... 196
  - 13.3.10 Near-term emphasis.......... 198
  - 13.3.11 The special case of startups.......... 200
  - 13.3.12 Avoiding procedure generation.......... 201
  - 13.3.13 Inertia 202
  - 13.3.14 Other obstacles.......... 203
- *13.4 Overcoming Obstacles.......... 203*

CHAPTER 14 HOW TO UPGRADE TO FORMAL DEVELOPMENT 206

- *14.1 How to Get There From Here.......... 206*
  - 14.1.1 New product startup 206
  - 14.1.2 Existing product enhancement.......... 207
- *14.2 A Paradigm Shift Is Needed 208*
- *14.3 Establishing Development Procedures.......... 209*
- *14.4 Start with a Robust FS.......... 210*
- *14.5 Management's Role.......... 211*
  - 14.5.1 Understand formal principles and their implications 211
  - 14.5.2 Evaluate costs and benefits.......... 212
  - 14.5.3 Decide and commit to that decision.......... 212
  - 14.5.4 Endorse, teach, motivate, and enforce formal procedures 213

- 14.5.5 Keeping score..........216
- *14.6 Summary..........219*

BIBLIOGRAPHY..........221

Index..........223

1
INTRODUCTION

The development of a new product is a critical activity for many companies. If a new product can be sold profitably, it contributes to the growth of the company. If not, it consumes the company's resources as it is repeatedly modified to seek market acceptance. The efficiency of the product development process contributes greatly to the overall success or failure of many companies. Most companies can significantly improve their results by upgrading their product development methods. This book provides guidance for making such a transition.

1.1 Product Development Methods

Small companies typically conduct their development programs in an informal, hit-or-miss fashion, intuitively managing the process on a day-by-day basis. After agreement has been reached on the general nature of the desired new product, its design begins. The detailed features of the product evolve as side effects of implementation decisions. As market considerations are discovered, changes are made to the product's goals, leading to redesign. This unpredictable process leads to schedule and budget overruns, and produces products whose structure wasn't coherently planned, but evolved as requirements changed.

Researchers have studied various development processes in a wide range of industries over many years. They correlated development practices with the efficiency of the development process and the quality of the eventual products, to identify procedures that frequently led to successful, high-quality products. Teams of researchers codified their findings into various formal methodologies that describe the practices of the most successful development organizations. A number of general principles emerged from this work that apply to diverse industries and technologies. This book describes a generic formal development method that may be used by companies of any size to improve their development results. These general principles are explained and endorsed, without advocating any specific pre-packaged methodology. The numerous substantial advantages of formal methods are documented and quantified by citing surveys and case studies of companies that are upgrading their development methodologies.

The formal technique starts by generating a detailed functional specification of the product to be developed, and then letting that specification **drive** the product-related activities of the company throughout the product's life cycle. This is the meaning of the title of this book, ***Specification-Driven Product Development***. Each company can use formal principles to synthesize a development procedure appropriate to its culture, technology, and marketplace. This book provides detailed recommendations for upgrading the maturity of a firm's development procedures.

Contrasted to an informal approach, a formal development program requires a larger initial investment of time and money, but saves both time and money over the life of the product and yields higher product quality. The fundamental thesis of the formal approach is that the initial investment in dollars and days to generate the specification is repaid many times over during the remainder of the product's life. This isn't a conjecture; rather, it's a summary of the worldwide experience of a multitude of companies of various sizes engaged in a wide range of industries over many years. The basic mechanism leading to these large returns on the investment is well understood. The careful planning that is done in the beginning results in fewer expensive changes throughout the lifetime of the product, and raises the quality of the product.

1.2 DEFINITIONS OF TERMS

Throughout this book, the terms "informal" and "formal" are used to describe two classes of development methods. This convenient simplification serves to separate intuitive development processes from those that follow rigorous, disciplined procedures. These two classes are characterized in Chapters 2 and 3. Throughout this book, a contrast between the two classes is repeatedly made to emphasize their differences. After preliminary concepts have been explained, a precise definition of these terms is given in Section 11.6.5.

A variant of informal development must be addressed. Frequently customers who have grown tired of coping with the products that emerge from informal environments require their suppliers to employ formal development methodology. Instead of complying, many vendors continue using informal methods, and supplement them by imitating some superficial aspects of formal methods. Extra resources are expended, but very few benefits are obtained in this way. Such informal processes are termed "informal-plus" development methods in this book.

The process of starting with a product concept and ending with the release of the design of a manufacturable product is termed "product development", with the understanding that a Production/Sales Phase will continue its life cycle. In informal systems, the entire development is conducted as a single phase, while in formal systems development consists of a "Definition Phase" (when the product's functions are defined), followed by a "Design Phase" (when the defined functions are implemented).

Product development generally follows a cyclic pattern. It starts with the development of a new product that isn't derived from any product that the company currently sells. The new product is built "from the ground up", with an original architecture and unique assemblies and components. Let's call this the "basic product". After this product has been accepted by the market, a sequence of enhancements to the basic product is made, adding features requested by customers and expanding the product's market into new niches. Let's call these "derived products". When the aggregate of these enhancements has stretched the basic product to the limits of its original architecture, the cycle starts over with the development of the basic product of the next generation. The length of the cycle depends on the industry involved; typically, it lasts several years.

This book strongly advocates a transition from informal to formal methods. When such a significant change is made in an organization, there are always people who quickly adopt the new methods and advocate their use, and those who resist change and cling to their old methods. In the literature concerning procedure modification, the technical terms "champion" and "laggard" are applied to these two groups, respectively.

Acronyms enclosed in square brackets (such as [SEI2, p. 7]) refer the reader to a similarly labeled reference in the Bibliography.

1.3 Assumed Development Environment

This book directly addresses the case of a small company with a few dozen to a few hundred employees that define, design, manufacture, sell, and service some tangible product. Much of this material also applies to organizations that develop products for their internal use (not for sale to others), and to the development of services rather than physical products. The terminology differs, but many of the concepts and principles continue to apply to these other business models. Readers interested in developing services should mentally substitute "service" for "product" and make other appropriate translations to apply this book's concepts to their individual circumstances.

In the assumed corporate environment, the President/CEO and Board of Directors are termed "top management". They supervise the "executives", Vice Presidents of Marketing, Engineering, Test, Manufacturing (and others), whose departments perform those classical operations. In the Engineering Department are found systems engineers and design engineers who are supervised by "managers" who report to the VP of Engineering. An independent Quality Department is assumed. The focus of the book is on procedure rather than organization.

Typically there are multiple products under development simultaneously, while other mature products have completed their development and are being sold. For the purpose of clarity, the focus is on the initial development or modification of a single product.

1.4 OVERVIEW

Chapters 2 and 3 introduce the concepts of informal and formal development methods and describe their typical results. Chapter 4 starts by contrasting informal and formal development programs with respect to their generation, storage, and utilization of functional information, to illustrate their differences and show how a formal program solves some of the central problems of the informal approach by its use of the Functional Specification. Chapters 5 and 6 further characterize the Functional Specification by listing the items that should be present and those that should not. The motivation behind these choices is also presented, to demonstrate that these decisions are the result of a reasoning process rather than random prejudices.

Chapter 7 presents other documents typically found in a formal development environment. Chapter 8 describes a formal review process, with concentration on the review of a Functional Specification and design documents. Chapter 9 describes some helpful subdivisions of the Design Phase. Chapter 10 discusses some aspects of the company's Quality Program and the desirability of certification. Some well-known examples of formal development methodologies are discussed in Chapter 11. The extensive benefits of formal development are presented in Chapter 12, along with references to some surveys and case studies that document quantitative benefits that were achieved by real developers. Chapter 13 addresses the concerns leading to a corporate decision to upgrade an informal system to a formal one, and Chapter 14 provides guidance for making this transition.

The principal purposes of this book are:

- To explain a generic formal product development method,
- To identify the ways in which it differs from informal methods,
- To set forth the benefits of formality, and
- To help a company upgrade from informal to formal methods.

This book is intended as a text for a college course on product development methodology at the Junior, Senior, or Graduate level. There is ample material here for a three-credit course with about 45 contact hours. As an alternative, these hours could be packed into an intensive one-week short course for industry professionals. This book could also be used in a self-study program. The emphasis is on basic principles rather than specific technologies, resulting in an interdisciplinary course accessible to Engineering, Computer Science, Business Management, and Marketing students. Although each example necessarily involves a specific technology, the methods are applicable to products of many types, including mechanical, electromechanical, electronic, hardware-only, software-only, and combined hardware/software devices. With appropriate translation, these concepts also apply to the development of services.

2
INFORMAL DEVELOPMENT

To explore the differences between informal and formal development programs, let's characterize a generic informal environment. Our focus is on the processes that are used within the firm, rather than on its organization chart.

Informal development is widely practiced by small companies. Some general concept of a desired product is formed, and the company's designers start implementing their understanding of this product. They start the design immediately and work frantically, in order to finish as soon as possible. The designers anticipate that functional changes will occur during design, so they usually start with parts they believe won't be changed. These parts are likely to be basic, low-level functions, so the design is done in a bottom-up (rather than top-down) order.

Since the design work started as soon as the product was selected for development, the marketers have not had an opportunity to fully define the features that the marketplace will require this product to have. Therefore, features are discovered (usually by contact with prospective customers) while the design is being done. Streams of new features, changes to known features, and priority changes impact the design team as it works. Since functional changes tend to extend the development schedule, they are generally resisted by the designers. The designers argue that the proposed change would require them to tear out an element that has been designed and tested, and re-design that element to implement the changed function, adding to the time and cost of the development. The marketers argue that the proposed change is necessary to sell the product. Executive time is needed to review these arguments and make decisions on a case-by-case basis. When a new feature or a functional change is approved, time and expense are incurred to make the design change, integrate it with the rest of the product, and test it. In many cases, a given feature is changed several times, causing re-design, re-integration, and re-test of that part of the product on each occasion. The combined effect of numerous functional changes is to greatly extend the development schedule. Since the cost of the development is typically dominated by designers' salaries, extending the schedule also causes cost overruns.

Schedules for informal development are frequently based on events that aren't related to the development process: a working prototype is needed to demonstrate at a trade show on a fixed date, or initial product delivery is required in the next fiscal quarter. A development schedule is constructed by working backward from the required completion date. Surprisingly often, the resulting schedule is barely achievable under best-case assumptions, including the absence of functional changes during design. The inevitable changes cause significant, unexpected, unpredictable schedule slips, expanding the duration and cost of the project. On each day, it looks like the design will be complete next month, so the project is allowed to continue. In some cases, the final scope of the project is two or three times as large as the initial estimates of time and budget. If the actual scope had been known initially, the project would never have been approved.

Throughout the design, management's focus is almost exclusively on attaining the next near-term milestone at its scheduled time. After several schedule slips, this preoccupation with schedule becomes a pathologic fixation. Eventually management is willing to compromise the product's functionality in order to get something out the door. The product's features are whatever was believed to be working on shipment day. Since the product still isn't suitable for customer use, implementation continues indefinitely under the guise of "Continuing Engineering", "Sustaining Engineering", "Customer Support", "Product Enhancement", or "Maintenance".

In an informal environment, there are no corporate Quality Procedures requiring comprehensive documentation for each product, so only partial documentation is generated. A bulletized list of product features may be written early in the development, and some aspects of the product's architecture and design may be recorded during the design process. Generally these documents aren't complete, detailed, or accurate enough to have much value. In the rush to implement changes, documents usually aren't updated to include the new feature or change, so they gradually become obsolete; the product differs from its documents in some unknown ways.

Informally developed products are subject to a common set of problems. Such products are hard to maintain and modify, since their functions, architecture, and design aren't accurately documented. As the developers who possess this information leave the company, transfer to other teams, and gradually forget the details, this crucial information is lost. As new features are added, they may interact in unexpected ways with existing features. This is an earmark of a low-quality product.

Early in the design process, the product's architecture is selected to support the functions that are known at that time. As unanticipated changes are made, the fundamental architecture cannot be re-optimized as each new feature set is implemented. Although the architecture becomes less and less matched to the product as it evolves, there is never an opportunity to start over with an appropriate architecture. The basic structure becomes convoluted as features that are incompatible with the architecture are added. The technical term that describes such an incompatible mixture is "kludge".

These classical problems tend to repeat in various forms and with similar consequences as each product is developed. In the informal tradition, no analysis of the development **process** is ever conducted; no attempt is made to trace the causes of problems to weaknesses in the development process, or to improve the process to minimize the effects of the problems. Each new development project repeats the steps that were followed in the previous project, even though the results of that previous project may have been quite unsatisfactory. Resources continue to be wasted, slowly producing products of low quality.

In one company, a Project Leader made a presentation of the status of his ongoing development project to his firm's employees. He started by boasting that his team had "completed" the development of the new software product in just 33 days. During the eight months thereafter, his team had been adding features to make the product acceptable to its prospective customers. At present, any feature can be made to work by itself, but when certain combinations of features are tried simultaneously, their implementations interact destructively in mysterious ways, and some of the features don't work. Debugging and correcting this class of unexpected problems is difficult and slow, so no reliable estimate of the delivery date can be made. In fact, it is not yet certain that the final feature list has been established. None of the developers or managers in the audience suggested that any of these problems might be the result of a poor development process.

Why are informal methods so widely used? It could be that development managers are unaware that more powerful alternatives exist. It could be that informal methods are simply easier to administer; there are no firm rules to constrain a manager's behavior, and every decision is made in an *ad hoc* manner. It could be that informal methods were used by a startup company and have been continued by inertia.

If you haven't witnessed informal development in person, you can watch an example on cable television. A development team constructs and demonstrates a single prototype, with no consideration for ease of modification, maintenance, cost of volume manufacture, development process, or documentation. The only constraint is an absolute time limit on the prototype phase of the development task. The name of the program is "Junkyard Wars". :-)

3
FORMAL DEVELOPMENT

Formal product development methods arose to correct the deficiencies in informal processes that are described in Chapter 2. People observed a number of iterations of the informal method and noticed that similar problems kept recurring. They gradually realized that these problems were being caused by the development methods being used, and devised remedies to improve the efficiency of the process and reduce the impact of its shortcomings. These improvements were collected into various aggregations to generate formal methodologies. Although these remedies were derived by different researchers in a number of industries in different countries and for various kinds of products, a number of formal techniques have been found to be helpful in a wide variety of environments. These common elements make up the generic formal development method described in this book. The fact that these same solutions were found in a wide range of situations establishes their validity and argues for their universal applicability.

In many cases, obvious cause-and-effect mechanisms link a formal process with one or more benefits that result from following that formal process instead of some informal alternative. In other cases, we observe that a particular benefit frequently occurs when a certain process is performed, and not so frequently otherwise, but no straightforward causal link is readily apparent. There is a statistical **correlation** between the process and the benefit, but we don't yet completely understand why they are related.

3.1 A HISTORIC PARALLEL

In the 1840's, Dr. Simmelweis observed that when doctors washed their hands between examinations, their maternity patients had a significantly higher survival rate. This correlation was sufficient to motivate some doctors to wash, even though there was no known mechanism that linked hand washing (a process) with patient survival (a benefit). Some other doctors refused to wash until a causal mechanism was revealed by the development of germ theory in 1870.

We can adopt formal methodology in our company "on faith", based on the correlations observed by numerous other firms. Or we can search for mechanisms whereby formality reduces or prevents many of the recurring problems we are experiencing. When this investigation is done with an open mind, often it's easy to discover causal paths from formal techniques to their benefits, since the deficiencies of informal development are a lot larger and more obvious than bacteria.

3.2 DEVELOPMENT PHASES

As described in Chapter 2, informal development consists of a single design step. The **fundamental characteristic that distinguishes formal methods** is that the development work is separated into **two distinct phases:**

- The **Definition Phase**—Decide what detailed functions a successful product provides and write them down.
- The **Design Phase**—Implement the functions you defined.

Clearly, if both steps are performed correctly, a successful product must result. Neither step is easy. Some of the assumptions hidden below the surface are that you are capable of defining the functions that are essential to product success, and that you possess the determination to carry out the development of the specified product without making significant changes.

3.3 Sequence of Phases

These phases must be done **sequentially**. The order in which tasks are done is as important as the activities performed. It is particularly important that the Design Phase doesn't start before management agrees that the Definition Phase is complete. It's very tempting to try to accelerate the project by overlapping the two phases, but doing this causes the process to degenerate to an informal one, with all of the associated problems.

3.4 Formal Documentation

A document called the Functional Specification is used to record the product's functions that are agreed upon in the Definition Phase. This document serves as an information conduit to deliver these functional requirements to the Design Phase. It acts as an interface between the two phases, decoupling their activities into modular components, allowing the developers to focus on one task at a time. The Functional Specification is clearly the cornerstone of formal methods. Chapters 4, 5, and 6 are dedicated to this document. Other essential documents are covered in Chapter 7. After these documents are written, they are subjected to individual formal reviews as discussed in Chapter 8, modified if necessary, approved, and released to the company's document control system.

Once released, these documents **drive** the activities that follow. The developers refer to them constantly, and include references to earlier documents as new documents are written, allowing functional requirements to be traced from implementation documents back to the Functional Specification. This process ensures that the objective of the Design Phase is achieved: the product precisely performs the functions defined in the Functional Specification, because every design activity is performed in response to a written functional requirement. **Driving the design from documentation** is the factor that differentiates a formal method from an "informal-plus" method that goes through the motions of generating some documents but doesn't use them to guide the following steps.

It's informative to consider the parallel situation in the construction industry. Do builders rely on conceptual sketches of the desired appearance of the structure, or do they prepare blueprints showing the details of the building's structure before breaking ground? Do they construct the building according to their blueprints, or do they discard the blueprints and build according to whim?

Formal development proceeds from Functional Specification to architectural design to high-level design to detailed design. Thus, a top-down order is followed: decisions that have the widest scope are make first, then decisions affecting a more limited scope, and finally decisions about details. As each decision is made, the underlying framework of prerequisites is already in place. One way to ensure that top-down principles are followed is to subdivide the Design Phase into several sub-phases that are done in a prescribed sequence. This matter is discussed further in Chapter 9.

3.5 FUNCTIONAL CHANGES

Formal methods provide adequate time during the Definition Phase to discover which features are desired by prospective customers, define them in detail, evaluate alternatives, resolve conflicting opinions, and agree upon the functionality of the product. Then the Design Phase produces only this **one** product. Features don't need to be added and continuously modified, as in the informal environment, because these alternatives have been evaluated during the Definition Phase and either included or excluded from the product then. This is a source of the formal method's great power. The design, integration, and test of a single product can be done much more quickly than the design, integration, test, reconsideration, removal, re-design, re-integration, and re-test of numerous features that is characteristic of informal processes. If each feature is changed twice on the average, enough development time and effort to develop three products has been expended, but only one product is available for sale. Since the desired functionality is known throughout the Design Phase in a formal environment, a robust, cohesive, high-quality design is possible. The scope of the design task is know in advance, so an accurate schedule can be constructed at the start of the Design Phase. Management must be continually vigilant during the Design Phase to permit only those functional changes that are absolutely essential to product success. If unnecessary changes are allowed to creep in, the design process quickly degenerates to an informal one, and the benefits of formality are lost.

In formal development environments, management's focus is on generating a robust Functional Specification during the Definition Phase, and on ensuring that the defined functionality is implemented during the Design Phase. Although timely completion is important, the emphasis is on achieving functional performance; development projects are feature-driven. This is quite different from the informal method's fixation on the near-term schedule at the expense of functionality; development projects are schedule-driven. Thus, an informal method always uses a direct approach to the current milestone; a formal method may use an indirect approach by doing other tasks that are believed to lead to success. As a quick analogy, suppose our task is to boil a pot of water as quickly as possible. An informal manager's attention is on monitoring the formation of tiny bubbles in the water. A formal manager concentrates on adding fuel to the fire under the pot, trusting his method to produce the desired result.

When the status report on the "33-Day Miracle" product (see Chapter 2) was presented, informal development had consumed nine months, with no end in sight. If the first month had been spent on determining the product's functional requirements, and the second month on deriving an architecture capable of supporting all features simultaneously before starting the detailed design, a much higher-quality product would have been delivered long ago.

3.6 Formal Methods Are Counterintuitive

In an informal environment, there is no underlying theory to guide development decisions, so they are made intuitively. If a task appears to advance the project toward completion of its nearest-term milestone, that task is started. Most choices seem so obvious that they just happen, without the managers being aware of any conscious decision process. If it is recognized that a decision is being made, the decision is regarded as a "no-brainer"; one path is so obviously correct that it is automatically chosen without any analysis. However, when one's brain is disabled, very poor decisions are likely to follow. The term "no-brainer" thus serves as a red flag.

If a formal method resulted in the same decision path (set of choices) that was made by the informal system, no benefit could occur. Therefore, the formal method must occasionally require a choice that differs from the path that would have been taken if this project were being conducted informally. Since informal decisions are intuitive, these different decisions must be **counterintuitive**. These differing decisions appear "wrong" to a manager whose previous experience was entirely informal. Such a manager instinctively resists following the "wrong" path, and seeks to "correct" the output of the formal decision-making process to align it with the intuitively "right" informal answer. It is in precisely these cases that the formal procedures must be most forcefully applied, since it is here that the benefits of formality are generated.

For example, suppose our formal method tells us that we must spend the next few days preparing to review an architecture document, then hold a review meeting (then probably amend the document and repeat the distribution, preparation, and review steps), and await formal signoff of the architecture document before beginning the design. An informal method would skip these steps and begin the design immediately, so as to complete it as soon as possible. Since we are (as always) in a desperate hurry, it's very tempting to try to find some excuse to follow the intuitively appealing informal path. It takes discipline and a firm belief in the superiority of formal methods to insist that approval of the architecture document must precede the start of the design activities.

Although "extra" time is spent now, the resulting increase in employees' knowledge about the product's organization and structure, the opportunity to correct defects and make high-level improvements before design commitments are made, the availability of a robust architecture document to drive the design activities and serve as a reference for design reviews, improvement in product quality, as well as other tangible and intangible benefits, probably accelerates the development enough to more than recover the "extra" time. Managers must have sufficient confidence in formal methodology to take the counterintuitive path, based on correlations with large future benefits, and avoid the intuitive path that provides smaller short-term benefits with certainty, but foregoes the future benefits.

3.7 BOWER'S INEQUALITY

After a company has been operating informally for a while, frequently an attempt is made to change to a formal method. A common misunderstanding is that the transition can be done by adding documents while continuing to pursue informal development. Certainly documentation is essential to a formal method, but equally certainly the other factors discussed in this chapter are also necessary to securing the benefits of a formal program. If documents are written but not used to drive the design, the effort to generate them is largely wasted. The following expression is used to emphasize that the addition of documents to an informal process doesn't yield a formal method. This expression also motivates the use of the term "informal-plus" to describe such a method.

$$\text{Informal Method} + \text{Documents} \neq \text{Formal Method}$$

3.8 Bower's Equation

If adding documents doesn't suffice, what expression can be used to define a formal method? Consider the following equation:

$$\text{Generate Excellent Documents} + \text{Follow Those Documents} = \text{Formal Method}$$

This expression emphasizes that documents are generated first and then followed (since they can't be followed before they exist). Note that the words "Informal Method" don't appear anywhere in the equation. This observation serves as a reminder that an informal process can't be augmented or modified to yield a formal method; it must be replaced. Note also that formal methods work with "excellent" documents that can be produced in the real world; unattainable "perfect" documents aren't needed.

4

THE FUNCTIONAL SPECIFICATION

4.1 STORAGE OF PRODUCT FUNCTIONAL INFORMATION

Let's start by considering what is meant by "functional" information. A functional specification defines WHAT the product does, as seen by its users. The product is considered as a "black box": its external interfaces are defined, but its internal workings are deliberately hidden. This is contrasted with an architectural, design, or implementation specification that describes HOW the product works. This important distinction is explored in Chapters 5 and 6.

One characteristic of an informal program is that the full functional definition of the product isn't available in any coherent written form. Instead, the information that defines the functions of the product is found in the memories of several individuals, supplemented by memos, emails, working notes, test results, meeting minutes, and abbreviated feature lists. These descriptions of the product's functions are typically mixed with product architecture and implementation details. Communication is often done by verbal exchanges, without making a permanent record of the topics discussed or the decisions reached. The product's functions change from time to time as new ideas emerge and implementation difficulties are discovered. The amorphous nature of the product's definition makes it difficult to keep track of the changes, or indeed to determine whether or not a change has actually occurred.

4.2 ACCESS TO FUNCTIONAL INFORMATION

Various employees of the company need to access this functional information. **Figure 1** contains an Information Flow Diagram showing this process. Groups of people are shown as rectangles, while circles represent documents. The sources of functional information are shown at the top of the diagram, and the users of that information are shown at the bottom. (A given person may belong to both groups.) The arrows show the flow of information concerning product functions from sources to users of information, either spontaneously or resulting from inquiries from users.

4.3 INFORMATION ACCESS PROBLEMS

It's fair to say that many small companies handle their functional information in this fashion. There are several inherent problems with this structure. The cumulative effect of these problems presents a substantial cost to the organization. In addition to the direct loss of person-hours and calendar delay, there are many intangible costs that are hard to identify and quantify, but may be more important than the obvious direct costs.

Let's examine in greater detail how functional information is handled in informal environments to see where weaknesses emerge, before discussing an effective solution to these problems.

Figure 1. Functional Information Flow in an Informal Process

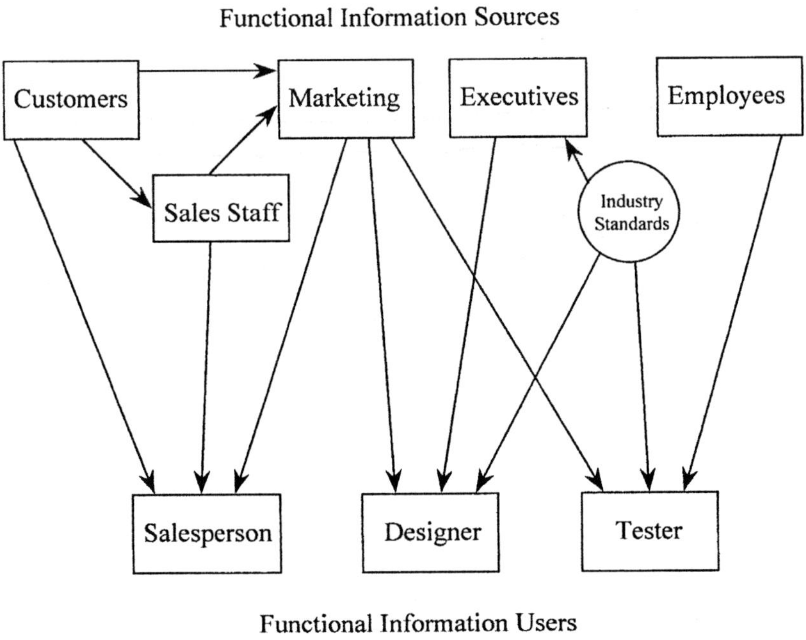

Functional Information Sources

[Diagram: Boxes for Customers, Marketing, Executives, Employees at top; Sales Staff and Industry Standards in middle; Salesperson, Designer, Tester at bottom, with arrows connecting sources to users]

Functional Information Users

4.3.1 Constant research is needed by many people

Many employees need to know what functions a product provides on a frequent basis. For example, a designer needs to know what features need to be implemented, and a salesperson needs to determine whether a given product meets a customer's needs. Whenever a functional question is raised, seekers of information begin a research process to attempt to find an answer. Information users may not think of their daily activities as "research", but they follow an informal program of investigation nonetheless. They identify sources of information likely to know something about the topic in question, formulate questions to those sources, collect replies, weigh those replies, and reach a decision.

When it's important to have current information (for example, a sale hinges on availability of a particular feature), the information user must repeat his research effort frequently, since features come and go without any formal notice that changes are occurring. The need for detailed functional information doesn't end when a product's development is complete, but continues as long as the product is being sold and serviced.

4.3.2 Repeated research may yield different answers

A particular question may be investigated several times, either because of the need to maintain current knowledge of a feature, or because two information users who both need the same data may be unaware of each others' investigations. A slightly different process may be followed each time a given question is researched. A different set of information sources may be identified, and a different subset of them may respond to the different queries. Their information and opinions may evolve over time.

Even in the ideal case, where identical answers are received by two information seekers, the decisions they derive from these answers may well differ. Each seeker is likely to weight information from his department's managers more heavily than that from a different department. Information users are likely to have different biases that influence their decision; they may favor results that make their own jobs easier.

All of these factors tend to cause different employees to reach different answers to common questions. The result is chaos. Salespeople promise features to customers that the designers have decided not to include. Features are developed and released to customers without having been tested in any way. Effort is spent designing product functions that no customer wants. Some of the difficulties caused by differing understanding of functional information are eventually discovered; their correction requires unscheduled time and effort, and may inconvenience customers. Other such difficulties may not ever be discovered. They continue to weaken the product and all products derived from it throughout their lives.

4.3.3 Decisions are delayed

Possessors of solid product functional information are likely to be busy people. When an information seeker approaches a knowledgeable source, a prompt response is rare. The source is out of town, on the phone, in a meeting, busy, ill, or on vacation. Appointments are scheduled and postponed. When the seeker finally gets the source's attention, the source is likely to refer the seeker to another information source, and the waiting cycle starts over. All too often, there are fundamental differences between the opinions of two or more information sources. In such a case, the information seeker may schedule a meeting some days in the future with the sources to settle the issue. If no mutually agreeable solution can be worked out, it may be necessary to escalate the decision to the executive level by scheduling another meeting. As a result, projects are lengthened while issues are being resolved.

4.3.4 Resources are consumed

The seeker's time is used to carefully formulate questions and attempt to contact sources. Sources spend time replying to these questions, and may need to do research of their own before doing so. If meetings are needed to resolve conflicts, additional person-hours are consumed by managers and executives. It's clear that enormous quantities of resources are being wasted by this informal process, and that the waste continues as the process is repeated throughout the product's life.

4.4 THE COMMON-SENSE SOLUTION

An obvious solution to all of these difficulties is to research each feature once, write down the results, make these results available in a library to those needing functional information, and update this material when changes occur. This approach is shown in **Figure 2**. When this process is made systematic, it leads directly to the formal documentation described in the following section.

Figure 2. Functional Information Stored in a Library

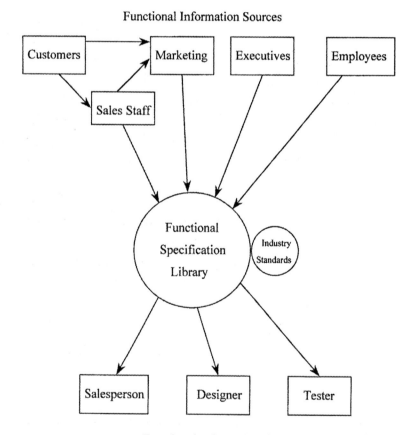

4.5 THE FUNDAMENTAL RULE

The written document that defines the product's functions is known by various names, such as Functional Specification, Product Specification, and Requirements. In this book, this document is called the **Functional Specification** (FS), since this is the natural name for something that specifies the functions of a product. It contains information that has been contributed by all of the departments of the company, and is intended for use throughout the enterprise and by certain outsiders. The following definition embodies a principle called the Fundamental Rule:

> **Fundamental Rule**
>
> **The Functional Specification is the only repository of product functional information.**

The definition of the FS is deceptively simple. It doesn't say that the FS is the "most important" information storehouse or the "best" of several repositories. Instead, it is an unconditional demand that no other reservoirs of functional requirements are to be considered. This absolute step is taken to solve at one stroke all of the problems that arise in informal systems because of multiple versions of the product's functionality. The remainder of this chapter addresses the creation and maintenance of the FS, the consequences and implications of the Fundamental Rule, the techniques used to ensure that the FS is unique, and the benefits that ensue.

4.6 GENERATION OF THE FUNCTIONAL SPECIFICATION

The Functional Specification is generated during the Definition Phase at the beginning of the development process. The process of deciding upon the new product's functions and recording them in detail in the FS is diagrammed in **Figure 3** and described in this section.

For each product, one or more employees or consultants are designated by management as the authors of the FS for the product. They collect product concepts and functional information from all sources, formulate a list of candidate functions and features, carefully weigh alternatives, make decisions, and record those decisions as a formal specification of the product's required functionality in a draft FS. This document is carefully studied and evaluated by a group of experienced reviewers, to ensure that it defines a product that will be desired by customers when it becomes available for sale. As the reviewers find errors and deficiencies in the draft FS, the FS authors make corrections and improvements, and the review process is repeated. Eventually a complete, accurate, robust FS that is satisfactory to the FR reviewers is produced, approved, and released for use. The generation of the FS must be done carefully, because this document will drive the remainder of the development effort.

The Definition Phase is the single time during the product's life that new features or functional changes can be introduced at low cost. It is the phase of maximum creativity and intellectual contribution. This unique opportunity must not be wasted by trying to shorten the duration of this phase, or by overlapping it with activities that should occur during the Design Phase to follow.

Figure 3. Functional Information Flow During Definition Phase

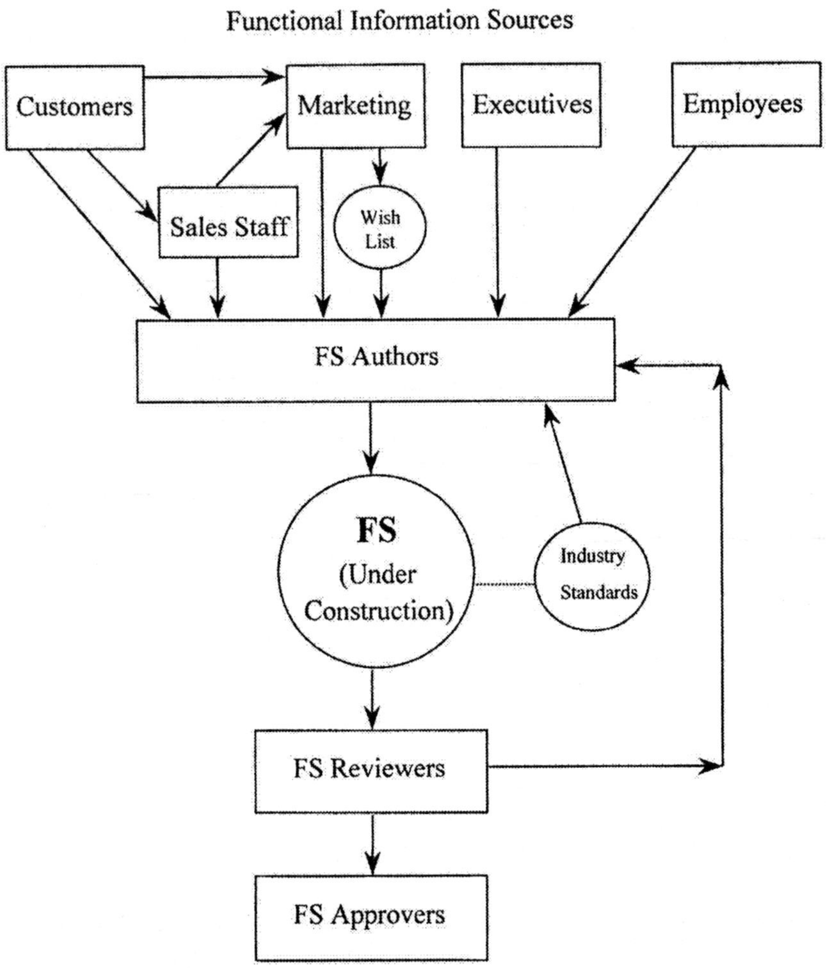

Functional Information Sources

```
Customers → Marketing    Executives    Employees
    ↓    ↘     ↓
    Sales Staff   Wish List
         ↓
       FS Authors  ←─────────┐
         ↓                    │
       FS                     │
    (Under  ···· Industry     │
    Construction)  Standards  │
         ↓                    │
      FS Reviewers ───────────┘
         ↓
      FS Approvers
```

4.6.1 Marketing contributions

The formal process concentrates the bulk of the marketing effort into the Definition Phase, instead of distributing it over all of the product's phases. The goal is to harvest all existing market knowledge and generate whatever additional market information is needed to decide on the product's detailed functions before the Design Phase begins. Customer wishes are weighed against development complexity during working meetings and reviews during the Definition Phase, and decisions are made and included in the FS as it is written. In this way, functional issues are resolved early, so that these issues don't linger and lead to marketing-driven changes during the Design Phase. Section 4.10.1 discusses the adjustments that the marketers make in their transition from an informal program to a formal one.

In order to focus the spotlight on the Marketing Department's contribution to this process, it is recommended that the marketers write a Wish List. The Wish List serves as a conduit for marketing to supply its distillation of customers' needs to the group that is composing the FS, typically the System Engineers. When the FS is completed, the Wish List has done its job, so it is discarded. Since the Wish List performs only this temporary role, it need not be a formally controlled document. Only its content matters; the authors of the Wish List don't have to follow any rules regarding document format or formal language. This absence of rules allows the Wish List authors to concentrate on producing a coherent description of the wishes of the significant customers for the product. The Wish List isn't a first draft of the FS; the Wish List represents marketing's understanding of the composite needs of prospective customers, independent of the realities of implementing the desired functions.

Preparing a Wish List doesn't discharge marketing's responsibility. Significant effort is required of the Marketing Department thereafter to contribute toward generation of the FS. The candidate functions and features collected from all sources must be reconciled with one another and evaluated. The functions that are selected for inclusion in the product must be described in formal specification language and reviewed. Although the primary author of the FS is typically a systems engineer, Marketing must contribute heavily to the execution of these tasks to ensure that the resulting FS represents a product that will be desired by customers.

The central purpose of the Definition Phase is to force functional decisions to be made, after careful review of the available pertinent information. If there are unresolved functional issues, the FS is incomplete, so the project isn't ready to leave the Definition Phase. Rules are stated in Sections 4.10.2, 5.2, and 6.9 to determine the level of detail that should be provided in the FS.

4.6.2 Roles of other groups

Although marketing carries the brunt of the effort during the Definition Phase, other departments must also make significant contributions. Engineers who have successfully developed similar products generally have a good understanding of customer needs, and have excellent ideas for features to satisfy those needs. It is essential to encourage them to exercise their creativity to the fullest to invent beneficial features, then capture their insights and embed them into the product during the Definition Phase. Test Department and Customer Service personnel are also aware of customers' needs, as well as the desirability of built-in operational status displays and diagnostic features. The Production Department may need test points or self-test features to expedite the production test process. All such inputs of functional requirements should be solicited for consideration, combined with concepts from others, and evaluated for inclusion in the FS during the Definition Phase.

The designers who will implement the product must be heavily involved in the Definition Phase. Their primary task is to perform feasibility studies of candidate features as they are proposed for inclusion in the product. The designers must ensure that it is feasible to implement the product defined by the FS [NASA1, p. 7]. The implementation should be planned in sufficient detail to determine the feasibility of the product without actually doing the design itself. Automobile mechanics and home repair contractors perform equivalent operations.

An experienced estimator can think through a repair project by visualizing the work to be done at each step, and make a reasonably accurate estimate of the labor and materials that will be needed, without actually performing any of the repair work. Product designers can learn to operate at this level.

The objective of the feasibility effort is to identify a design approach capable of satisfying the FS requirements with a high degree of confidence. No attempt is made to do a detailed design, to polish this approach, develop alternatives, or select the "best" approach; a single compliant design concept proves feasibility. When the designers become aware of the totality of the functions that they must implement, they may well optimize their implementation by choosing design approaches that differ from those that were used to establish feasibility.

Since the FS isn't approved until the end of the Definition Phase, the final requirements aren't known, so the actual design process must not be allowed to start; no formal design documentation should be produced during this phase. However, the designers should keep careful informal notes of their architectural assumptions and tentative implementation choices, to serve as input to the Design Phase to follow. [The designers also estimate the development and production costs of the product for inclusion in a Business Plan; these tasks are outside the scope of this book.]

4.6.3 Sources of functional information

In addition to the contributions from the various departments discussed in the previous sections, all other sources of functional ideas should be actively sought. Trade shows and technical and trade publications may disclose competitors' features. National and international standards provide recommendations for the details of features, as well as interfaces necessary for interoperability with established products. Marketing reports concerning industry segments can be purchased, and marketing consultants can be hired to generate recommendations for the product's functions. The company should consider all available information sources, select the sources to be used, and make them available in the early part of the Definition Phase.

4.6.4 FS authors

Employees from all departments may serve as authors and/or reviewers of the FS. If the product is being designed for a specific customer, a representative of that customer may participate as an author. In other cases, a consultant who specializes in writing FSs may be retained. It doesn't really matter who writes the FS; everyone's word processor and graphics package are more than adequate for the job. The reviewers actually determine the contents of the FS, by repeatedly requesting changes to the FS until they are satisfied with the result. If the authors are familiar with the principles presented in Chapters 4, 5, and 6, fewer iterations are needed.

The principal authors of the FS are typically systems engineers. Perhaps this is because the Engineering Department is the heaviest user of the FS immediately after its approval; the engineers want to expedite the Definition Phase so they can start the next phase, Design. They also want to ensure that the FS contains enough detail to lead to an orderly development.

4.6.5 FS reviewers

An interdisciplinary team of technical and marketing experts and managers is chosen to review the FS. These reviewers study and discuss drafts of the FS as they are produced by the FS authors, and provide specific feedback to the authors. Features evolve as a result of invention, refinement, and negotiation. The FS reviewers weigh their estimates of the benefit of a given feature against the cost of its development, and decide whether the feature is to be included in the FS. The content, organization, formal language, and level of detail of the FS are examined and refined. Chapters 5 and 6 give suggestions regarding items to include and avoid in the FS.

A danger is that the FS authors may not recognize the value of a desirable proposed feature and decide to leave it out of the draft FS. The FS reviewers aren't aware of this situation because the feature isn't mentioned in the text they are reviewing. A solution is to provide some overlap between the authors and the reviewers.

When general agreement is reached among the reviewers that the FS is complete and that all open issues have been resolved, a formal review meeting is typically held. Its purposes are to pick up any omissions and errors, ensure that full agreement has been reached, and present the FS to the executives and managers who will approve it. Minor problems cause adjustments to the FS before it is submitted for approval. If major problems or open issues are discovered at the review, the Definition Phase continues; the problems are solved, the FS is revised accordingly, and the review meeting is repeated. Chapter 8 covers techniques for reviews, with specific information on reviewing the FS in Section 8.5.

4.7 APPROVAL OF THE FS

The FS is a crucial document, since it will determine the activities of many people throughout the company, as well as its vendors and customers, for years to come. After it has successfully completed its formal review meeting, it is submitted to management for approval. The FS should be carefully reviewed and signed by management, to ensure that it describes the desired product, and to give it the authority that may be needed to persuade employees unfamiliar with formal development programs to rely on it. It should be signed by the head of each department (usually a Vice-President) and then by the President/Chief Executive.

4.8 CHANGES TO THE FS

After the FS is approved, there must be formal methods in place to distribute it to authorized users, prevent unauthorized changes, and to allow controlled changes to occur.

4.8.1 Why are changes bad?

Although a functional change presumably improves the product, the change itself always disrupts the orderly development or production of the product. Except in the rare case of a change that reduces the scope of a product by simplifying or deleting a feature, the change delays the product's release or delivery while increasing its cost.

Let's treat changes that occur during the product's development first. A wealth of experience reveals that functional changes during the development cycle are the principal causes of poor designs and cost and schedule overruns. Whenever you see a world-class kludge, it's almost certain that several functional changes occurred late in the development process. Some projects never recover; they are abandoned as total failures.

When the FS is first approved, it is distributed to people throughout the company. Many of these people start using the FS to perform their jobs, as described in some detail in Chapter 12. They base their work on the FS because their employer has assured them that the FS describes the product that will be built and sold. Test plans, manuals, and brochures are written, designers are transferred or hired, architectures and designs are derived, parts and equipment are ordered, prototypes are fabricated, and specific features are promised to customers. Whenever the FS changes, some of this work may need to be repeated. If this happens often, employees will lose confidence in the formal method, which will appear to them as being the same as its informal predecessor. In self-defense, they will delay their use of the FS of each future product until it has "settled down", causing some of the benefits of the formal method to be lost. This loss of employee confidence is intangible, but its importance shouldn't be underestimated.

A significant change to an electronic product may require printed circuit boards to be redesigned. Several weeks are generally needed to get the new boards fabricated. If prototypes have been built, the boards that are being replaced must be discarded as scrap. Once integrated circuits are soldered into printed circuit boards, it's not usually feasible to remove them for transfer to the new boards, so these components must be scrapped as well. Replacement components are ordered. For certain "long lead time" components, the replacements may not arrive for months, causing a substantial schedule slip to the entire development project.

The impact of a software change may not be as spectacular as that of a hardware change, but is no less destructive. A real-life example illustrates an important problem.

The Saga of the Flashing Indicator

A simple electronic controller was being developed for a client. It was based on a primitive microprocessor, running a serial software architecture. As input events arrived, the processor would quickly process each transaction to completion and then await the arrival of the next input.

Late in the development schedule, the client added a new requirement: when a particular error condition was detected, an LED indicator was to flash. This appears to be a trivial change. The processor executes a single instruction to turn the LED on, delays a while, then executes another single instruction to turn the LED off. What could be easier?

It turned out to be surprisingly hard to implement the delay. The LED had to stay on for a minimum time to be visible. The processor couldn't wait idly, timing this delay, because other inputs needed attention. There was no real-time operating system to schedule a task to start at a specific later time to turn off the LED. A solution was found, but the resulting increase in complexity weakened the product, consuming system resources and reducing the safety margin whereby the other requirements were being met. These factors also reduced the ability of the design to accommodate future changes, and consumed development time.

The fundamental problem was that the new requirement violated the assumption that each task could be completely processed before the next task was started. The architecture that resulted from this assumption was fine for the original features, but poor for implementing the unexpected new feature. If the flashing requirement had been identified before the design began, a more flexible architecture could have been chosen to smoothly incorporate this feature into a robust structure.

It has been observed that <u>design modifications are **never** thoroughly checked</u>. Several factors prevent <u>design changes from being checked to the same degree that the original design was validated</u>. The most obvious cause is that the change is always done in a hurry, to avoid or minimize a schedule slip. In a few hours, the modifier attempts to re-create the design steps that may have taken several people months to accomplish, to verify that the original assumptions are still valid after the change. The assumptions leading to design decisions probably weren't fully recorded. Some choices were made intuitively; the designer wasn't aware that a decision process was taking place. Under these conditions, the modifier cannot possibly trace the thought processes that led to the original design. The original design may have been reviewed at several levels by a group of knowledgeable people; the modifier has no such advantage.

Another factor works against the modifier. It can be explained in terms of levels of abstraction. A group of atoms forms a semiconductor junction that is part of a transistor that is part of a flip-flop that is part of a register in the arithmetic unit of a microprocessor that is part of a personal computer that supports a word processing application that allows this sentence to be typed, contributing to your understanding of this concept. As each level is made to work, it serves as a building block to implement the next higher level. It is essential to make use of the objects and relationships at each level as conceptual units, abstracting their high-level properties while ignoring the details at all lower levels. It would be impossible to implement a word processor if the behavior of the individual atoms had to be considered at each instant, without intermediate levels of organization.

Let's return to the topic of design. As each level is debugged, the designer's mind eagerly leaps to the next level of abstraction. When a design modification is done, the modifier must retreat some unknown number of levels, to the lowest level where an assumption may need to be verified. It's a difficult intellectual process to give back all of those hard-won levels and start over with fundamentals. The modifier's tendency is to strongly resist a full retreat, and compromise on a review at a higher level than is appropriate.

For example, it is well known that software modification is an error-prone endeavor. A comprehensive study considered software changes to high level language modules to correct specific defects. It found that the changes to fully one-half of the modified modules contained errors. In some cases, the original defect wasn't repaired; in other cases, new problems were introduced by the change.

Changes can also occur to a product after its development is complete and the product is in volume production. Such changes can be quite expensive, because they may cause a production run to be scrapped. If the change is being made to correct a serious defect, products that have been delivered to customers may need to be upgraded or retrofitted.

4.8.2 Why are changes necessary?

Given the high cost associated with changing a product, it would be wonderful if all changes could be prohibited. However, it's unreasonable to suppose that the original FS defines the perfect product, or that market conditions won't change over the product's life. When a competitor makes an unexpected announcement of a desirable new feature, to preserve market share it may be necessary to introduce a corresponding feature in our company's products.

4.8.3 Change control

After spending the time and effort to build and approve the FS, a document control system is used to preserve its contents, make it readily available to all authorized users, and allow necessary modifications while protecting it from unauthorized changes. When it is necessary to make a functional change to the product (either during its Design Phase or after it is in production), the FS is carefully updated to define the new functionality. This change to the FS causes changes to the design documents, and then to the product's design. The process whereby the FS is changed is diagrammed in **Figure 4** and described in the following sections.

4.8.4 The Modification Request (MR)

When a functional change is wanted, a document called a Modification Request (MR) is written to carefully define the desired operation. (MRs are also used for notification of implementation errors and other purposes; the emphasis here is on MRs that apply to the FS.) The MR should contain a detailed description of the desired function or change, so that the MR reviewers can evaluate its merits and costs without initiating research investigations. An MR that simply states a problem or symptom may need significant exploration before the best functional change can be clearly stated. While it's important to expedite the MR process, decisions should be deferred until the functional details are fully formulated and the feasibility and cost of their implementation have been determined.

38   Specification-Driven Product Development

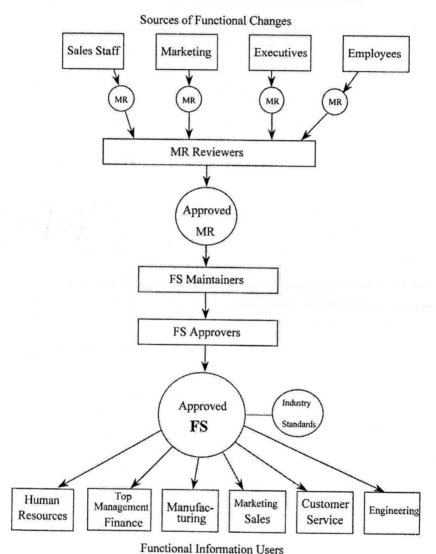

Figure 4. Functional Information Flow After Definition Phase

4.8.5 Review of MRs

A group of managers is typically assembled to review accumulated MRs on a periodic basis. These MR reviewers perform a critically important function. They are the gatekeepers who are charged by top management with maintaining the integrity of the FS, the central element of the formal development process. The MR reviewers must have the experience and maturity to recognize the substantial company-wide impact of **any** functional change, including intangible costs like employee morale and confidence in the formal method. They need the authority and dedication to reject all but the most essential changes. They must put aside their natural desire to make the perfect product, in favor of reinforcing the formal development method, particularly during the time that the company is making the transition from an informal system.

If the MR reviewers approve capricious changes, the addition of luxury features, or changes to cater to the whim of a single possible customer (who wants the color of the standard product changed to match his draperies), employees quickly recognize that management isn't committed to the principles underlying a formal method. If the FS is constantly changing, it can't be trusted to represent the final product, so its value is drastically reduced; the time and labor whose costs are stored in the FS are wasted. The employee realizes that the so-called formal system is the same as the old informal system, except that he is expected to run the spelling checker on his word processor before submitting documents.

4.8.6 FS modification

Figure 4 shows some FS maintainers who accept MRs approved by the MR reviewers and convert them into modifications to the FS text. Of course, the groups of FS authors, FS reviewers, MR reviewers, and FS maintainers may overlap. The skills required of the FS maintainers are largely the same as the FS authors. They must preserve the integrity of the FS by applying the same principles that the original authors used. These principles are presented in Chapters 5 and 6, and Section 8.5.

As the FS is modified, the text that defines a given requirement should retain the same section number, because other documents refer to the section numbers of the original FS. If these numbers are preserved, the other documents may not need to be updated. If a requirement is completely removed, notation such as "This section intentionally left blank" or "The requirement on viscosity was removed in Revision D" can be used as a placeholder to allow the following sections to keep their original numbers.

4.8.7 Re-approval of the modified FS

If extensive changes have been made to the FS, the MR reviewers can inspect the changes to verify that they reflect the MR reviewers' intention. The modified version of the FS produced by the FS maintainers is re-approved by the same senior executives who signed the original FS. (If lower-level personnel were allowed to approve changes, inappropriate content could be introduced, and the authority of the FS would be undermined.) Once the new version has been approved, it completely replaces the original FS as the governing document. An announcement is made to alert the FS users that they must start using the new version.

4.9 VIOLATIONS OF THE FUNDAMENTAL RULE

The principle that the FS is the only repository of functional information is easy to state but surprisingly difficult to implement. This section details some of the ways that the Fundamental Rule might be compromised. In some cases, a well-intentioned employee strays from compliance, while in other cases the rule is deliberately broken. This section is included as a warning, because violations can be insidious, easy to carelessly commit, difficult to detect, and quite detrimental to the success of the project. Executives and managers must exercise continual vigilance to be sure they are always following the Fundamental Rule themselves, and to detect and correct any departures by others before they cause harm to the development project or the product. This section contains a lengthy list of types of violations, but many other varieties are possible.

4.9.1 Sneak paths

A "sneak path" is any route other than the approved FS whereby functional information reaches an information user. The flow of information may be initiated by its source; for example, a manager's unauthorized directive may waive a specific FS section. The flow may also be instigated by its recipient, an information user who asks a functional question.

Clearly, any sneak path is a violation of the Fundamental Rule. If a sneak path is allowed to exist, all functional information users must make use of it, if they are to receive the same information as those who do. They must all compare the sneak path's output with the FS and appeal to Higher Authority when differences are found. They are right back to the informal system. Frequent attempts to form sneak paths may indicate that employees don't fully understand or accept the Fundamental Rule, or that the FS is incomplete. If the formal system doesn't allow employees to obtain the functional information they need to do their jobs, informal alternatives will arise to fill the gap.

Let's look at some specific sneak paths to see how they arise. **Figure 5** shows a number of typical sneak paths leading from various disallowed sources of functional information to the users thereof. (Of course, any user may try to get information from any disallowed source, not just the ones shown in this figure.) There is only one allowed source of functional information, the most recently approved version of the FS. Consultation of the FS is indicated as the Approved Path. All information users must consult the current FS to obtain answers to their functional questions, while disregarding all other sources of functional information. This is a very simple rule, but considerable practice is required by practitioners of informal systems before compliance becomes automatic.

The sneak path from the Industry Standards in **Figure 5** requires explanation. It represents the use of part of a standard not specifically identified as a requirement in the FS. Designers sometimes imagine that all sections of a standard must be met, whereas only a subset is intended. Salespersons like to expand the stated requirements to include entire standards, to simplify the explanations they make to prospective customers. This matter is discussed further in Section 5.4.

Figure 5. Sneak Paths

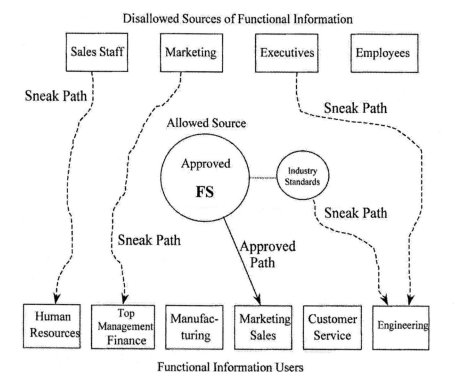

4.9.2 Managerial functional decisions

Suppose a designer approaches his supervisor and says, "I'm about to start designing the Control Panel, so I need to know whether or not to include a Reset button. What should I do?" The supervisor might respond, "We talked about that at the Product Planning Meeting. Let me see what Mary's latest thinking is, and I'll get back to you with a decision."

This manager has spoken two sentences, and has made four mistakes. First, he indicated that the Product Planning Meeting is a source of functional information. Second, he named Mary as an alternate information source. Third, by offering to render a decision on a functional matter, he has established himself as an information repository. Fourth, he has missed what is clearly a badly-needed opportunity to remind himself and the designer where the answer is to be found: **the FS**. The sneak paths generated by his response are shown in **Figure 6**.

It's natural for a manager to want to help a subordinate resolve functional issues so that development may proceed. How can a manager do this without generating sneak paths? The manager must firmly direct the subordinate toward the FS, and nowhere else(!), each time a functional question arises. The manager may say, "I'll be happy to help you understand and interpret the FS, but as you know, I can't modify or waive any part of it. Let's pull the on-line FS into my word processor, and search it for 'Reset' and 'button'...Aha, here it is, hidden in the Control Panel Section." To encourage FS use, the manager may give subtle hints: "I noticed your copy of the FS isn't nearly as wrinkled, smudged, and dog-eared as mine. Are you on this project full time?" After a few such prompts, the subordinate will come to understand that his next performance review will be a lot smoother if he starts phrasing his functional questions in terms of specific FS section numbers. If other subordinates have the same difficulty, more training on this point is indicated.

Figure 6. Managerial Functional Decisions

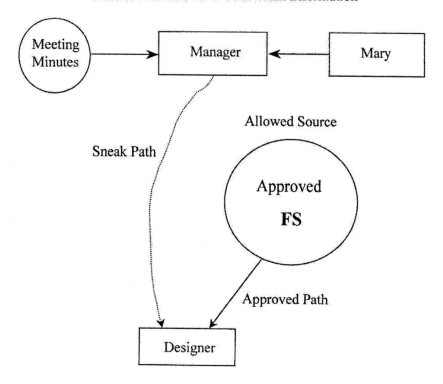

4.9.3 Parallel FS and Wish List

Another kind of sneak path occurs when some other reference document is deliberately substituted for the FS as the controlling authority. Consider a marketing manager (let's call her Susan) who wants to make a clandestine change to a product being developed according to an approved FS. The formal MR avenue isn't attractive, because the MR reviewers have been asking questions like, "Is this proposed change driven by a recent, unforeseeable shift in the marketplace, or is it due to lack of attention during the Definition Phase?"

Susan decides to use the Wish List as a vehicle to do an end-run around the formal procedures. She resurrects the old Wish List that had been used to help generate the FS. She modifies the Wish List by adding the feature she overlooked, then wheedles signatures from her marketing colleagues. "I'm just trying to keep our internal marketing notes up to date. These executives have already reviewed and signed it. Approval is only a silly formality."

As soon as the ink dries on the last signature, Susan does an abrupt about-face. She waves the newly-approved Wish List at the designers, demanding that her new feature be immediately added to the product. "This Wish List is co-equal with the FS. It sets forth the all-important Market Requirements; it's What The Customer Wants, so it has special status. These recent signatures show that all these executives agree that the Wish List governs in this case. We're just bringing to the surface what has always been implicit in the Wish List, so it's not a valid reason for slipping the delivery schedule. You can change your FS to include this feature, if that will make you feel better."

Figure 7 shows the situation that Susan is trying to bring about. The Wish List and FS stand as functional authorities of equal rank. Marketing is allowed to use their Wish List as a royal road to sneak unreviewed functional changes into the product, while other departments must use the MR-FS route for their changes. Users of functional information must consult both authorities, and conduct research to resolve differences; they are right back to the informal system.

Clearly, this is an absolutely intolerable situation. By violating the Fundamental Rule, the benefits of the formal system are lost, while its costs remain. It also makes a farce of the entire Quality Program, by circumventing the most crucial review process.

Figure 7. Parallel FS and Wish List

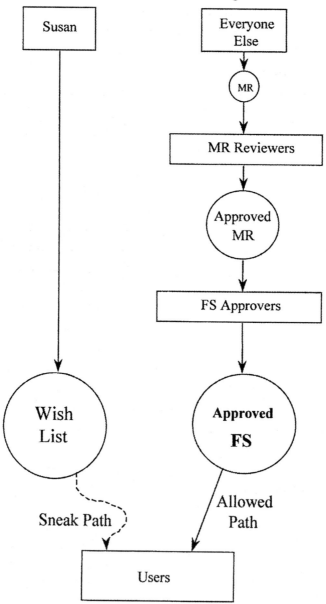

To prevent anything vaguely similar to this from ever being attempted, the company-wide Product Development Procedures must make it completely clear that the Wish List is alive only during the Definition Phase; it totally vanishes and has no further function after the first version of the FS is formally approved. There is no point in amending the Wish List thereafter, since it has no purpose or validity. Asking executives to approve a meaningless document should alert them that a deliberate process violation is being attempted. Having the status of the Wish List clearly defined in published corporate Quality Procedures gives the company ample grounds to terminate Susan's employment for her egregious behavior.

4.9.4 Collected FS changes

Another way that competing functional information storehouses arise is in the administration of the MR process. To illustrate this, let's briefly trace an FS change chronologically through its evolution.

At first, the change is conceptual. Someone gets an idea for improving the product, or works out a way to remove some defect. A proposed change is formulated and discussed with others. When informal verbal agreement is reached, an MR is written and submitted to the MR reviewers. The next step is the approval of the MR. The change is then incorporated into a new version of the FS by its maintainers, and the modified FS is circulated for signatures. The newly-approved FS is then distributed or posted, an announcement is made, and the new FS takes effect.

All these steps take time. As a change progresses from initial concept to approved document, it becomes more and more likely that it will become effective. At which step should employees start re-directing their work toward the new goal? The official answer is that workers always follow the most recently approved FS, instantly re-orienting their efforts when a newly approved version is announced [DUNN, p. 149; YOURDON, p. 45]. In practice, it's unreasonable to continue following an obsolete FS when the effort of doing so is wasted; for example, continuing to develop a feature that is about to be removed from the FS. We want to avoid wasted effort, yet we cannot allow "grapevine" rumors of impending changes to become a rival storehouse of functional information, as shown in **Figure 8**.

Management can take two steps to relieve this conflict. As a practical compromise, an employee working in the area of a pending change can be directed to stop work that would be wasted if the change were approved, and spend time on unrelated tasks instead. This isn't a violation of the Fundamental Rule, since no functional departure is involved. All management has done is to rearrange the order in which design tasks are to be performed. Once the new FS has been approved, the suspended task can be officially revised or removed from the Development Plan.

The second way that management can solve this problem is to expedite each of the change steps recited above. The MR reviewers can either meet frequently, or they can meet promptly when a significant MR is submitted. These reviewers can perform their due diligence activities promptly, then act decisively to accept or reject the proposed change. The new FS can be "walked around" to solicit signatures, instead of loitering in executives' in-baskets. Signature sheets can be sent to executives in parallel, instead of requiring all to sign the same sheet in serial fashion.

One step that management cannot take is to "unofficially" direct employees to begin work toward the pending change. This is a blatant violation of the Fundamental Rule; the manager's directive becomes a second source of functional information.

Figure 8. Accumulated FS Changes

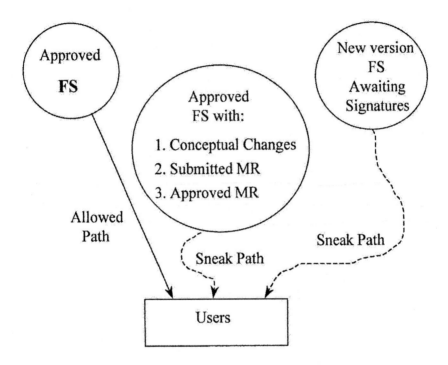

The practice of accumulating changes to the FS until a "critical mass" warrants formal revision must be avoided for two reasons. First, slowing the approval cycle is exactly what we have just been trying so hard to avoid. Second, the fact that frequent changes are occurring is a symptom that the process isn't working properly, most likely due to inadequate attention to the Definition Phase. This crucial symptom must not be concealed from the executives. If they are annoyed by being asked to review and approve frequent FS revisions, they are likely to provide the badly-needed feedback to ensure that the Definition Phase of future projects will receive the concentrated attention it requires. Since many of the benefits of the formal approach aren't obvious when its costs are being paid, every such source of constructive feedback must be exploited.

4.9.5 Evolutionary development

In an informal environment, the product's specific features are crafted by its designers during the development process. Starting from general functional goals, the designers invent implementation techniques on the fly, and the details of the features fall out as results of this design process.

When designers who are experienced with informal techniques start working in a formal environment, it's natural for them to bring their informal methodology with them. They feel free to modify the features defined in the FS to fit their preferred implementation techniques. If an alternate function is easier to design or test, and if they can convince themselves that some customers may prefer it, the change is made without pursuing the formal MR route or notifying anyone.

Another designer may depart from the FS by attempting to generalize the product, to make it easier to add features of a particular type in the future. If she incorrectly guesses the direction of those features, the functions she adds may not be helpful, and may in fact consume the resources needed by the new feature. The ability to expand a product is desirable, but shouldn't be added on a freelance basis; with a little thought, it can be quantified and put into the Architecture Specification, as discussed in Section 7.3.

A similar phenomenon occurs at the end of the design cycle. Unless a developer is forcibly stopped, he will continue to endlessly refine the product, making it ever more elaborate and wonderful.

On the other hand, when a developer feels that the allowed time is inadequate for him to complete all of his assigned tasks, he may prioritize these tasks and put no effort toward the task he perceives as least important, in order to complete the others. If the developer isn't sure how to implement the feature, or if it involves a lot of tedious, unpleasant work, this may color his evaluation of its importance to the customer. By ambiguously reporting the status of the ignored task, the developer may escape detection until late in the development schedule. The developer "wins" if he can stall until the deadline is reached and the ignored feature must be postponed until the next release.

The result of these phenomena is "feature creep", the gradual migration of the product away from the FS. The equivalent sneak paths are shown in **Figure 9**.

Such a rogue developer may argue that he is improving the product, but his unilateral changes are actually quite destructive. A developer would rightly complain if some "external" agent were to require functional changes that impacted his "internal" design to be made from time to time, without warning or authorization. The developer must be led to understand that many other people throughout the company are using the FS during the development cycle in order to gain the benefits listed in Chapter 12. They are depending on the FS as a static definition of what the finished product will do. [At worst, they operate with the knowledge that the MR reviewers are (in principle) taking into account the impact on their work of any proposed change before approving it, and that they will be informed of approved change.] These people are taking irreversible steps based on the FS, such as purchasing components and equipment, and (most important) making commitments to customers that specific features will be available. To each of these people, the rogue developer is the "external" agent whose arbitrary changes (when discovered) are causing them to do unplanned rework to their individual "internal" operations and undermining their commitments.

Even though functional improvements aren't being sought after the Definition Phase ends, some good ideas are certain to emerge during design. These ideas are processed as MRs, becoming formal FS enhancements or being added to the Future Feature List for possible inclusion in a later release.

Figure 9. Designer's Specification Sources

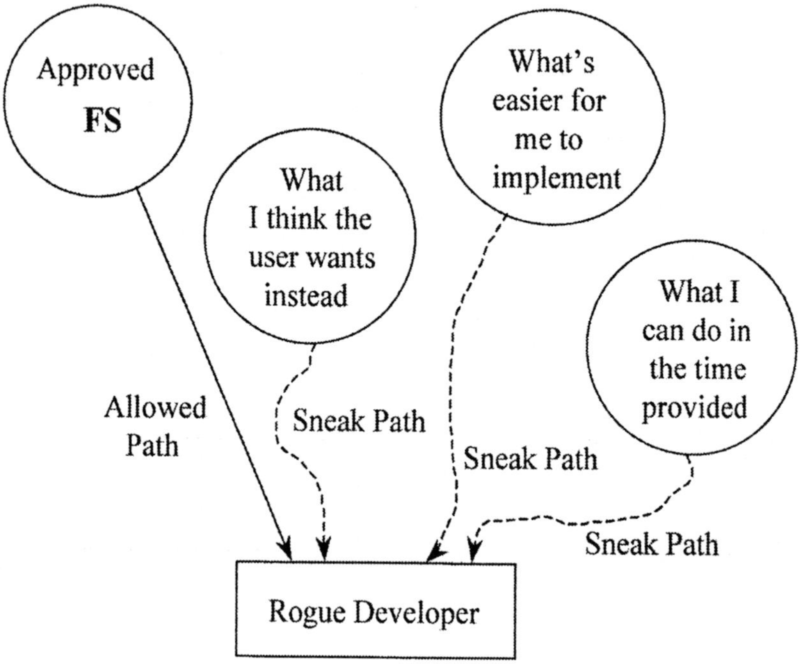

```
   Approved              What's
     FS                 easier for
                        me to
              What      implement
              I think the
              user wants          What I
              instead             can do in
                                  the time
   Allowed                        provided
   Path      Sneak Path
                        Sneak Path
                                    Sneak Path
                    ↓
              Rogue Developer
```

4.9.6 Inter-departmental leaks

Can you spot the sneak path in **Figure 10**? Hint: it's small, and near the bottom of the figure. The line between the Designer and the Tester represents a leak of functional information between their departments that didn't come through the FS. Over lunch, the designer told the tester about the interpretation the design group had devised for a particular troublesome FS section. The tester will base his Acceptance Test Plan on that interpretation, ensuring that the product as designed will be accepted.

Figure 10. Inter-Departmental Information Leak

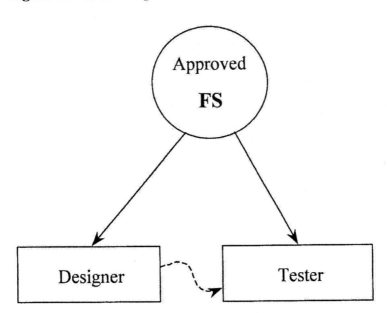

This seems normal enough, doesn't it? What's the problem? The problem is that the test department is supposed to serve as an independent "check and balance" on the designers. Their purpose is compromised if they pick up the designers' interpretation instead of making their own study of the FS. Testers must not be influenced by the designers' preconceptions about the product's functions [NASA1, pp. 9-10].

In addition, if this FS section is so vague that it is subject to significantly different interpretations, won't all of the other users of the FS be confused also? Since the designers are uncertain of the meaning of an FS section, they should write an MR to clarify that section according to some defensible interpretation they prefer. This gives the FS reviewers and maintainers a shot at examining and correcting the language of the FS. Thus, the uncertainty is resolved before time is wasted, and the quality of the document is increased, benefiting all its users.

4.10 OTHER PROBLEM AREAS

Other difficulties can arise even though the Fundamental Rule is carefully followed. In many cases, these problem areas can be anticipated, so that management can intervene proactively.

4.10.1 Winning employee cooperation

It's natural for employees who have been working in an informal environment to want to continue using the methods they have learned. The thrust of the formal program is to **replace** the sub-optimum and sometimes destructive procedures that have evolved into the informal program with formal techniques that have proven beneficial to numerous other firms over many years. To make the transition, the company must overcome this inertia and win the employees' enthusiastic cooperation with the new methods.

Let's examine the changes in the nature of the work performed by various groups of employees as they move from an informal to a formal environment.

The importance of participation by **management** in the formal program cannot be overestimated [SEI3, p. 23]. In informal organizations, executives perform day-to-day management, constantly making low-level decisions and solving mundane problems. In a formal system, subordinate employees follow the established procedures to perform much of this low-level work themselves. The executives are no longer managing directly; they are now exerting their authority indirectly, by way of the Procedures they have established. Section 14.5.4 contains suggestions of management activities that motivate employee cooperation.

Marketing personnel have to make a significant change in their working methods as a transition to formality is made. In an informal system, the marketer collects user requirements over the life of the product, as information drifts in from various sources. The evolving requirements cause a continuous stream of functional changes to the product. Marketers are notorious for talking to a single prospective customer, then requiring a standard product to include all of the quirks, special features, and unique gimmicks that customer said he wanted. The product is whatever is needed by the customer who called most recently.

The major thrust of a formal program is to aggressively seek functional requirements during the Definition Phase, and then **stop** this activity when the FS is completed. The principal goal is to reduce the number of changes to the product after its design begins; most of the benefits from a formal program flow from this factor alone. The marketing effort that was spread over a period of years is concentrated into a few months. Marketing personnel work on products serially, concentrating their efforts on a single product during its Definition Phase, rather than working on multiple products in parallel through the products' lives.

Finding out what your customers really want is hard. Finding out what your customers will want next year, in order to start developing it now, is very hard. Finding out what the people who will become your new customers next year will want is very, very hard. This is the key task the Marketing Department must perform. It is crucial to the success of the whole enterprise. Management should recognize that marketing has changed its methods and greatly increased the intensity of its information-gathering efforts, and provide suitable rewards for successful performance. It should be noted that the Marketing Department receives a large share of the benefits of the formal program, as detailed in Section 12.5.

The **designers** of the new product must also make a transition in their methods. Starting from the informal techniques described in Chapter 2, the designers need to separate their functional creativity from their implementation creativity, performing the former in the Definition Phase and the latter in the Design Phase thereafter. That is, they must rigorously follow the FS, allowing it to drive their design activities. They must exercise discipline to implement exactly WHAT the FS says, and nothing else, instead of inventing and modifying the features as they design.

When a transition to a formal development program is announced, designers read the new Product Development Procedure and typically complain that the formal system will force them to do "extra work". The basis of this belief is that the documentation effort is **additive** to the informal work, which will be continued and won't benefit from the documents (see Section 13.2.1). Clearly, the designer doesn't understand that the relationship (Bower's Inequality) stated in Section 3.7 is an inequality. Since the familiar informal process is being **replaced** (not augmented), we can't know without experience whether the new method will require more effort or less effort.

Some firms employ informal design methods, while writing an FS in an attempt to gain the benefits of (or give the appearance of) using a formal approach; we are calling this an "informal-plus" approach. Definers and designers start at the same time and work concurrently. The designers make architectural choices and high-level design decisions early in the project. As desired features emerge from the definers, designers naturally try to reject features that are incompatible with their architecture or are inconvenient to implement within their design framework. After a few such projects, designers become adept at arguing against functional changes. If they can quickly implement and validate some low-level functions, or copy part of the design of an earlier product, they can make the claim that these portions of the design are "done and tested", so that any minor functional change will force a redesign and a lengthy re-test, slipping the development schedule by many weeks. The cost of repeating the test of the entire product is attributed to each proposed change. This argument has been repeated so often that it has become a single word, "doneandtested".

By quickly establishing a "doneandtested" beachhead, and by haggling over every inconvenient feature in every FS draft, designers can delay the release of the FS until their design is nearly complete. Then management is forced to accept the design as the functional reference, since it is too late to make any significant change. Thus, the implementation must be done in a bottom-up fashion, accommodating the interfaces of the initially "doneandtested" components, since it has been argued that they cannot be changed. The definers are reduced to following the designers around, documenting the side effects of the design decisions as the product's functions; all consideration of what the customers want has been lost. The informal designers have achieved their goal: there is no functional reference for their work, since the Acceptance Test is based on the FS, which is written after the fact to describe whatever was designed. The design cannot be wrong or incomplete, since it is its own reference.

Another factor is also at work here. Some designers instinctively try to delay the FS release to avoid personal responsibility for schedule overruns. "You can't expect us to deliver in June; the FS wasn't even finished until April." The conflict between definers and designers makes the entire development slower and more costly than either a purely informal or a purely formal approach.

Competent executives cannot tolerate such deliberately wasteful behavior. By making just one procedural change (requiring that the FS is released before design **begins**), management can reverse the designers' destructive motivation. They now want to expedite FS release, so that they can start the design task.

As shown in Chapter 12, many **other groups of employees** benefit from a robust FS. For the most part, they get a free ride; they don't have to create the FS or modify their work habits much to take advantage of this new resource. They benefit from getting earlier, more detailed, and more reliable knowledge of the product's functions, so they can start their tasks earlier in the development cycle. Since changes are minimized, less rework is required.

4.10.2 Continual refinements

Suppose the FS generation process bogs down, spending a lot more time than management allocated to the Definition Phase. The product's functions are continually refined, decisions are made and unmade, different FS organizations are being tried, and more and more details are being added.

This is normal and healthy. It's much cheaper to make functional changes while the "product" is a text file in a computer than at any later time. In fact, one study found that the cost of correcting a functional problem after delivery was 100 times greater than correcting it during the specification phase. If the Definition Phase takes a long time, that indicates the product's functions were poorly formulated at the beginning; substantial value is being added by the Definition Phase in this case. Conceptual experimentation with product functions leads to higher product quality, if only by rejecting bad ideas.

One way to keep the FS generation focused and productive is to find a leader for this phase who has had broad experience developing functional specifications for other products. Such people are available as consultants.

Of course, it's possible to allow the Definition Phase to extend past the point of diminishing returns, producing a "wheels within wheels" elaboration of little incremental value. The basic cutoff rule appears in Section 5.2: if a feature could possibly be implemented so badly that it would impact customer acceptance, it's important and needs to be in the FS. Others don't.

Management can limit an unproductive Definition Phase by establishing a near-term completion date. Generally, this is done too early rather than too late, and the price is paid in frequent later changes.

4.10.3 If agreement can't be reached

It may not be possible to reach agreement on fundamentals between individuals or departments. Engineering can't design the functions that

Marketing believes the market requires, while Manufacturing is unable to fabricate or test what Engineering wants to make. If such an impasse can't be resolved, the conclusion is easy: the product shouldn't be built.

In an informal system, each department might approve development of the product in general, not realizing that a conflict exists, or expecting to "fix" the design to conform to its wishes before development ends. In a formal approach, trying to determine the specific features of the product in advance reveals that no mutually satisfactory product exists. This is an enormous benefit, both in saved development cost and opportunity cost; the time and money can be spent developing a different product, one with a chance of success.

4.10.4 The role of prototyping

"Prototyping" refers to a development method in which a portion of a product's functions are modeled in an exploratory manner. The prototype is used to demonstrate variations of these functions by adjusting the prototype's parameters. The advantages of this approach are that a variety of features can be rapidly studied, and the user can be shown a representation of certain features early in the development process.

This sounds frighteningly like the informal approach: don't specify anything, just head off in the general direction of a product, tinkering and iterating until the development schedule runs out. However, it is possible to reconcile prototyping with formal development methodology. Let's see what can go wrong and then examine remedies.

Suppose a prototype is implemented during the Definition Phase to study a visible function such as the user interface. To make the prototype's output more realistic, more and more of the product's functions are included during the Design Phase. No functional specifications or formal design documents are produced. The only "reviews" consist of demonstrations of the prototype's output.

The designers maintain that they are exempt from compliance with the company's Product Development Procedure. "We aren't developing a product. This is different. It's not a deliverable product at all, just an exploratory vehicle, a...uh...prototype! Yeah, that's the ticket!" During the Design Phase, the prototype becomes more elaborate, imitating most of the product's functionality. When the delivery date arrives, no other development has been done, so the only alternative is to ship the prototype to the customer. Abracadabra, the magic handkerchief is lifted, and the prototype has **become** the product. If management tolerates this deception, it has allowed the designers to totally circumvent the company's most important Procedure by using "anything goes" informal methods to generate an unspecified, undocumented, unreviewed, untested product. The fact that this was done quickly carries absolutely no weight with the Quality Department, which should prohibit delivery.

Let's consider a different case, in which prototyping is used during the Definition Phase to demonstrate an innovative feature to a prospective customer, who participates in fine-tuning the prototype until total satisfaction is reached. The customer then insists that the final product must be "exactly like" the optimized prototype. The vendor tries to address this demand by declaring that the product's functions are "what the FS says, plus whatever the prototype does". Of course, this is a blatant violation of the Fundamental Rule, since it establishes two sources of functional information. To cure this problem, the second try is: "The FS consists of this document plus whatever the prototype does". This is also unacceptable, because:

- It still doesn't correct the fundamental two-source issue, so all of the problems described in Section 4.3 above will plague the project,

- There may be contradictions or duplications between the document and the prototype (see Sections 6.2 and 6.6),

- It doesn't distinguish between the characteristics of the prototype that are important to the customer and its incidental characteristics, so every detail of the prototype must be re-created in the product (see Section 6.9); the tail is wagging the dog,

- For the same reason, the tester doesn't know what functions to test in the Acceptance Test Procedure,

- To achieve perfect imitation of the prototype, designers are tempted to smuggle undocumented fragments of the prototype into the product, and

- The prototype isn't as portable or reproducible as a document, so many of the benefits listed in Chapter 12 are diminished or lost.

How can the advantages of prototyping be obtained without sacrificing the basic principles of a formal system? If certain restrictions on prototyping are adopted and carefully followed, no conflicts need to arise.

Suppose prototyping is allowed only during the Definition Phase. Exploration can be done freely then to help determine the details of the product's functions. When the features emulated by the prototype have been optimized, the aspects of the prototype that are important to prospective customers are described in the FS, using sentences, tables, diagrams, figures, and pictures, as appropriate. There is no good reason to treat features determined by prototyping differently from features chosen by other means. (If part of the prototype's user interface consists of displays on a computer screen, software is available to "capture" an example of each unique screen into a file for inclusion into the FS, so these screens need not be manually entered.)

At the end of the Definition Phase, all functional decisions have been made, so there is no need for a vehicle to explore alternatives. Therefore, the prototype is required to be dismantled at the end of the Definition Phase. Note that this rule is precisely analogous to discontinuing the use of the Wish List at the same time, as described in Section 4.6.1. Both steps are taken for the same reason: to avoid any possibility of future conflicts with the FS. The reader is encouraged to re-read Section 4.9.3, substituting "prototype" for "Wish List".

NASA's formal procedures [NASA1, p. 14] contain equivalent provisions concerning prototypes. Prototyping must be completed before implementation begins. If any element of the prototype is to become part of the final product, it must be fully documented and must pass all "checkpoints" (i.e., reviews, unit tests, and integration tests) that apply to the parts of the product that are being developed "from scratch".

By observing these reasonable precautions, prototyping can be effectively used to help determine functionality without compromising the integrity of the formal process.

4.11 OTHER BUSINESS MODELS

In the remainder of the book, it is assumed that a company determines what product features will be desired by prospective customers and internally generates an FS to capture this knowledge. The company has the ability to unilaterally modify this FS in any manner during its Design Phase. For the purposes of this section, let's refer to this situation as the "Internal" model.

There are other possible business models. A large customer may know exactly what it needs, write a detailed FS to record its requirements, and distribute this FS to potential vendors to solicit bids and technical proposals. A winning bidder is selected, negotiations take place to arrive at a mutually satisfactory FS, and this final FS is locked in as the cornerstone of a binding legal contract, calling for the vendor to develop a product complying with this FS and deliver it to the customer. Any departures from the FS must be negotiated between the vendor and customer, leading to an amendment to their contract. Let's call this situation a "Contract" model.

In another case, a vendor may work with a potential large customer to jointly agree upon the features of a new product. The two parties cooperate to author and review a FS that sets forth the desired features.

The customer may agree to fund part of the vendor's development cost, in exchange for the ability to "steer" the FS toward specific features that accommodate this customer's individual needs. The vendor usually retains the right to sell this product to other customers. After design starts, changes to the FS must be negotiated between the two parties; substantial changes may require increased development funding by the customer. Let's call this the "Hybrid" model.

When development is done formally, the differences among various business models are confined to the Definition Phase. Once an FS has been established, it serves as the vehicle to convey the customers' desired functions to the designers. The designers work in exactly the same way in all models: they implement the FS. By decoupling activities in the Definition Phase from those in the Design Phase, a formal methodology allows a vendor to develop several products simultaneously using more than one business model, to match the business models of customers of various kinds.

In order to keep the designers' activities uniform across business models, designers who work in the "Internal" model should behave as though they work in the "Contract" or "Hybrid" models, in which the FS has been fixed by contract with an external party. This customer will be very attentive to all FS features, requiring any non-complying features to be corrected and any unspecified features to be removed before acceptance. It will be very hard to get customer approval of any functional change for the convenience of the designer. When the designers adopt this viewpoint, they stop making freelance changes to the product's features and concentrate on implementing the FS as written. Designers who have had the experience of working in a "Contract" environment can easily migrate to a formal "Internal" or "Hybrid" environment, because they already have the correct attitude; their work is driven by the FS. As noted elsewhere, designers who have worked only informally often have difficulty moving to a formal environment, because they don't share this attitude.

5

WHAT'S IN A FUNCTIONAL SPECIFICATION?

In Chapter 12, dozens of different uses of the FS are presented. Some of them appear to have contradictory goals: the FS must be so specific that the product can be designed with no other functional information, yet it must not disclose sensitive company data to outsiders who read it. This could be achieved by maintaining different versions of the FS for various purposes, but this is exactly what is prohibited by the Fundamental Rule. How can we be sure that a single document will satisfy all of the users simultaneously? In this chapter and the following one, simple rules are set forth to ensure that the FS contains all of the information its users need, while avoiding inappropriate disclosures. If these rules are carefully followed, **a single FS is generated that is suitable for all purposes**. This is a surprising and happy result.

If a list of desirable components of the FS is given without explanation, the reader may interpret these elements as the prejudices of the author of this book, and have no strong reason to replace his or her own past practices with the listed components. To motivate the reader to upgrade his or her skills as an FS author, the reasons that other FS authors have found the recommended elements to be necessary are discussed, and the elements are related to specific benefits from Chapter 12. Examples of business situations that illustrate the reasons for inclusion are given in some cases.

The following ingredients should be included in each FS.

5.1 BOUNDARY DEFINITION

The extent, or boundary, of the product must be explicit in the FS. Think of drawing a box around the product, to separate the deliverable parts of the product (inside the box) from the customer's environment (outside the box). The FS authors must clearly state which elements are part of the product, and fully specify the functions they perform, while avoiding a description of the surrounding devices supplied by the customer. The practice of rigidly excluding all of the elements that are beyond the control of the product's designers allows them to focus on the functions their design must provide. The explicit boundary prevents confusion about which elements are to be provided, which helps the FS authors to maintain consistency throughout the document.

This boundary also serves to highlight the interfaces between the product and its environment. Since the FS is a "black box" definition of the product as seen by its user, it is crucial to specify all interfaces between product and user. The first step is to identify all such interfaces; they are the signals that cross the boundary. Care must be taken to include non-electrical signals, such as the light from an indicator that the user observes, the sound of an audible alarm, and the manipulation of switches. A picture that shows all of these interfaces crossing the product's boundary is called a "context diagram" [YOURDON, p. 157].

As an example, suppose we have decided that the electrical power cable and the plug at its end are to be included in our product. The product's boundary is between our plug and the connector into which the user inserts this plug. The FS must specify the physical characteristics of the plug; this is an important function, since the user must provide the mating connector. The FS must also detail the primary power that the user must provide, in terms of voltage and frequency ranges within which the product will operate, the maximum current that the product will require, as well as ripple, noise, and transient limits that the product will withstand.

All of these items are characteristics of the physical and electrical interface between product and user, and must be included in the FS. The authors of the FS may have excellent ideas about the wire gauge that the customer should use to provide power, and the grounding and fusing practices he should follow, but this material must not be included in the FS, since it doesn't describe the product. These suggestions may be provided to the user in an Application Note, but they are definitely out of place in the FS.

5.2 COMPLETE FUNCTIONAL REQUIREMENTS

The boundary analysis of the previous section reveals the product's input and output signals. In principle, there are mathematical "functions" that relate each output to all of the inputs. These relationships need not be derived and recorded in the FS, but the concept and terminology are the same as those a mathematician uses. The "functions" we seek to describe in the FS are the rules that define the behavior of each output signal in terms of the present and past input signals, for all possible combinations of present and past values of the inputs; that is, the transformations that generate the outputs from the inputs [NASA1, p. 7]. The FS is a complete specification of WHAT the product does, as seen by its user, without any description of the manner in which the product achieves this performance [YOURDON, p. 159].

Since the FS is the only source of functional information, its authors must be careful to ensure that all of the details **that are important to the user** are included, and that there are no unstated assumptions. Characteristics that aren't important to users should be omitted. To help make this determination, the following exercise is often helpful: if it is possible for the feature to be implemented so badly that it could affect customer acceptance of the product, it's important and should be specified in the FS. When deciding what to specify, the authors anticipate that the designers who work from the FS will follow two principles:

- If a feature isn't in the FS, it won't be in the product,
- Any unspecified aspect of a feature will be the result of the lowest-cost design choice, without regard for user convenience.

To ensure that a satisfactory product will emerge from the development process, the FS authors must include all relevant specifics of the desired product. The FS must answer all of the designers' functional questions [YOURDON, p. 162]. When numeric quantities are specified, their acceptable ranges, limits, and tolerances should be given.

5.3 Formal Language

An FS is a list of requirements. In English, requirements are stated by using the word "shall". "Shall" is used and understood by all professional writers and readers of functional specifications throughout the English-speaking world. Authors must always use "shall" in each sentence that states a functional requirement, and for no other purpose, because FS readers will interpret it in precisely that way.

A writer of informal "specs" might argue, "The ITU-T Recommendations don't use 'shall', so I don't have to." But that's precisely the point: they're Recommendations, quite different from Specifications, so they use different, and appropriate, language.

Some informal "specs" use a mixture of present and future tenses by occasionally using the word "will". The author may be unconsciously describing the true status of the features she is defining; features that have been developed and are presently available are described in the present tense, while features that haven't yet been developed are accurately described in the future tense. A poker player would caution the author not to reveal the cards in her hand so completely to an adversary. A customer reading the FS will realize that the "present" functions are really in the product, while the functions described in the future tense are hypothetical or subject to change or omission. It may be desirable to conceal this situation from the customer. This problem is avoided by consistently using "shall" to state requirements.

In addition to using proper language to express requirements, the FS must be precise, logical, and rigorous. For example, if the product has two binary inputs, does the FS fully define the product's performance in all four possible cases, or just the most likely three cases? For another example, if a feature is described as "optional", the intended meaning of this term must be clarified. Is one of the possible options permanently installed at the factory, like standard and automatic transmission "options" on an automobile? Are two options both present in the product, with only one available at a time, like AM and FM functions of an AM/FM radio? Or are both options to be available simultaneously, like "optional" car radios and air conditioners?

5.4 Reference to Available Standards

It's often important for products to comply with established interface standards, operational procedures, and conventions, in order to be compatible with other equipment that the customer may own. Rather than copying lengthy material from standards, it is good practice to reference standards documents in the FS. These standards must be available to all readers of the FS. Since some of these readers may not be employed by the company that wrote the FS, internal documents shouldn't be referenced. Rather, "open" public standards are used, such as those from IEEE, ANSI, Telcordia, and ITU-T.

The FS should reference specific sub-sections of standards, rather than the entire document. If a standard is referenced without qualification, every user of the FS must research the entire standard document and all of the documents it refers to, and all of the documents **they** refer to, and so forth, to find the applicable material. This is a great deal of work, which is avoided by identifying specific sub-sections of each standard with which the product must comply.

All of these standards include sub-section numbers before every few sentences. This isn't an accident. The authors of these standards know that those who use their standards need to refer to specific sub-sections.

FS language such as, "The product shall comply with the applicable sections of Standard X" is intolerable. This statement forces all serious readers of the FS to research the entire document tree and to decide individually which sub-sections apply. These readers are likely to reach different conclusions, in a way highly reminiscent of the informal approach. For example, suppose Standard A refers to Standard B, that in turn references Standard C. In order to ensure that the product meets two small sub-sections of Standard A, the FS authors incorrectly specify compliance with the entire Standard A. The writer of the Acceptance Test Plan researches the entire document tree, and finds the general statement, "These devices are intended to operate in the office environment", in an obscure sub-section of Standard C. This causes her to spend weeks characterizing the "office environment" and planning a series of environmental tests to ensure compliance. This effort was wasted, since no one else intended environmental tests to result from specifying Standard A.

There is a temptation for marketers to refer to whole standards, to make it easy to describe compliance to customers, and to give them the warm feeling that the product complies with "everything". However, promising generalized compliance allows the customer to imagine that the product meets his interpretation of sub-sections that the designers never considered. This is an invitation to disappointment and litigation. It is usually the systems engineer's role to restrict the claims of compliance to the sub-sections that are actually necessary.

Let's focus on the problem by asking, "Do we know which specific sub-sections apply?" If we know, all we need to do is write these sub-section numbers into the FS. If we don't know, then the reference to "applicable" sections is meaningless and deceptive; the authors should remove any mention of this standard from the FS.

5.5 Consistent Terminology

Names of functions and component elements of the product should be defined early in the FS, and should be used consistently throughout. If the FS mentions "efficiency", "utilization ratio", and "percent usage" in three sections to refer to a single number, the reader of the FS will understand that three different quantities are to be calculated. It is even more confusing to refer to different things by the same name. Surprisingly often, a major assembly will be given the same name as the entire product, leading to continual confusion.

5.6 Logical Organization

A major purpose of the FS is to communicate the customers' requirements to people unfamiliar with them. It should contain a brief introduction explaining the nature of the product and its environment, a statement of the scope of the FS, and definitions of terms. All of this material should be written without using the word "shall", to make it clear that no requirements are being stated.

The body of the FS, where the requirements are set forth, should be organized in a top-down fashion. That is, the principal functions should be presented first, followed by secondary functions and features of lesser importance. Each section of the FS should completely detail a given function; if the various aspects of a function are scattered among several FS sections, a laborious search is required, with the possibility that a relevant sentence is overlooked. If the use of several sections to define a single function can't be avoided, each section should cross-reference the others by number.

5.7 Numbered Sections

The FS should be divided into multiple sections, each of which ideally defines a single function or feature. The length of a section can range from a single sentence to a few paragraphs. The sections should be numbered in a hierarchical manner, as is done in this book. These section numbers are necessary to allow other documents, such as the design documentation and test specifications, to refer to specific parts of the FS. Some practitioners take this principle a step farther, numbering the individual requirements consecutively throughout the FS to distinguish them from the surrounding explanatory text.

5.8 The Product's Name

The name of the product permeates its documentation. Abbreviations and mnemonics derived from the product's name are used extensively in the names of documents, computer files, software variables, module names, and circuit assemblies. If development is started using a particular product name and then that name is changed, the documentation is thrown into chaos. It's not feasible to track down all abbreviations of the original name and replace them with equivalents based on the new name, because of the time needed. What always occurs is that a confusing hodgepodge of nomenclature based on a mixture of the two names evolves. This is a great nuisance that is easily avoided by choosing the final product name during the Definition Phase, and using it consistently throughout the FS.

It is generally the province of Marketing to choose the name under which a product is sold. This is a small part of the contribution of the Marketing Department to the Definition Phase. If Marketing is unable to choose a product's name, it is clear evidence that marketing research isn't nearly complete enough to finalize the FS. After half of the substantial effort required of Marketing during the Definition Phase has been expended, naming the product should be easy. This is a meaningful "litmus test" of the Marketing Department's commitment to formal methods. If Marketing wants to use an "internal" product name during development and then change to an "external" name for sales purposes, or suggests a change to the agreed product name after FS release, a significant problem has been revealed. Other methods should be used to achieve whatever goals Marketing is trying to meet by changing the product's name.

6

WHAT'S NOT IN A FUNCTIONAL SPECIFICATION?

An FS could contain all of the ingredients listed in Chapter 5, and still be made ineffective by the inclusion of destructive material. As with the desirable items, it's important to discuss the reasons for excluding certain items, and give examples of the danger of their presence. The following elements should **NOT** be included in an FS.

6.1 Anything that Isn't a Function of the Product

Since the FS lists product functions, it's pretty obvious that things that aren't functions of the product should be excluded. In addition to distracting the reader from the actual requirements, extraneous material is detrimental if it conflicts with requirements or produces confusion. The practice of rigidly limiting the FS to product functions has been found to be highly beneficial. Some common examples of foreign items that infiltrate functional specifications follow.

6.1.1 HOW the product works

Engineers are notorious for including implementation details when they write an FS. The FS tells WHAT the product does; it is the **input** to the design process, whose **output** is the design, telling HOW the product operates. The designer who works from the FS should be free to select the least costly design solution consistent with the functional requirements, rather than having her hands tied by a partial specification of the result of her effort.

As an example, consider an FS that defines a clothes dryer that has a time display and a keypad. Terms such as "computer", "processor", "program", and "code" don't belong in the FS because they describe a particular class of implementations; they aren't functions of the dryer. Even though you think a microprocessor will probably be used to handle the keypad and display, you should hide this information to make the FS independent of the design choices. Imagine a sequence of simpler and simpler dryers, ending with a primitive dryer with only an ON/OFF switch. At some point in this sequence, the processor disappears, since it's cheaper to implement very simple controls with discrete logic. For dryers near this processor/discrete boundary, it may not be clear in advance which technology will turn out to be cheaper. For those dryers, it wouldn't be prudent to commit to one approach or the other before this topic is explored during the Design Phase. Therefore, the FS should leave this decision open. By extension, the FS for a complex dryer should also avoid anticipating any result of the design process.

6.1.2 Wishes

Wishes are expressed by indefinite words like "should", "desirable", "optional", and "preferred". The technical literature on wishing makes it clear that wishes are granted only to possessors of magic lamps. Since wishes don't express requirements, **they will be ignored** in the design and test processes, and none of the "desired" features will be provided. Including such material serves only to embarrass the FS authors.

If the product is being developed by an external contractor on a fixed-price basis, you can be sure that all such "desired but not required" features will be omitted. (See Sections 12.7.11 and 12.7.12.)

If we measure the efficiency of our internal design staff by comparing its cost to an external contractor's bid, as suggested in Section 12.7.11, we shouldn't expect our employees to lower their score (and jeopardize their continued employment) by including features the contractor won't provide.

The authors of the FS must evaluate the cost of providing each feature and its value to the user, and then **make a firm decision** to include the feature in the product being specified, or to put it on the Future Feature List for possible inclusion in a later product generation. Avoiding this responsibility by using fuzzy language passes the buck to people less qualified to make this decision. Using imprecise language in an attempt to reserve the right to informally amend the FS by "reinterpreting" the vague language during the Design Phase without acknowledging that a change occurred is equivalent to the atrocities described in Section 12.1.4. A simple summary of the formal process might be: Postpone the start of the Design Phase until **all** functional decisions have been made, and then implement those decisions. If the first step isn't completed, the second step can't be done well.

6.1.3 Extraneous topics

The FS has a definite, specific purpose. It shouldn't serve as a catch-all for non-functional material such as development schedules, personnel assignments, cost estimates, and marketing plans. These items are important, but they don't describe functions of the product. They belong in Business Plans or Development Plans. See Sections 6.10 and 7.6.

6.1.4 Specification of the design process

So far, this section has used examples of things that aren't functions. Another class that should be excluded is things that are functions, but not functions of the **product**. For example, the phrase, "Consideration shall be given…", attempts to specify a function of the **design process**, not of the product. An experienced designer will cross out the remainder of the sentence without reading it, knowing that it won't generate a test item. Of course, this is exactly the opposite of the writer's intent.

Such phrases reveal a buck-passing gambit. Having recognized the need for further "consideration", the FS authors should perform this "consideration" themselves, formulate the result as a product functional requirement, then record it in the FS.

6.2 Duplication

On the surface, duplication seems to be a good idea. We explain a requirement, then explain it again in a slightly different way, in case the first explanation wasn't fully understood. If communication were the only purpose of the FS, this would be a good approach. However, the need for rigor overrides the desire to clarify the explanation via redundancy. The FS authors should combine their multiple explanations to produce a single coherent description. Suppose the FS has two descriptions of the same feature. To bring the issue into focus, let's ask, "Are the two descriptions identical?" If they are, one of them can be omitted without losing any information; the FS is shortened and simplified, too. If the two descriptions aren't identical, they are different. If they differ, one might include a requirement not mentioned in the other; in this case, the less detailed description can be omitted. The other possibility is that they describe a common feature differently. In this case, they are in conflict with one another, and represent a contradiction, which is discussed in Section 6.6.

A subtle trap exists when identical copies of a requirement are carried in the FS. If a change is made to this requirement, the FS modifier may not find all of the copies of the requirement. Then the copies that were changed conflict with the copies that weren't.

An FS may mention a function in its Table of Contents, again in the Introduction, again in the statement of requirements, and possibly in a summary. The use of formal specification language in the statement of requirements, and only there, allows the authors to make it clear that the requirement is being defined only once, and where that definition is located.

Suppose that a customer is given a copy of the FS, or a proposal written from it, containing two different descriptions of the product. As a simplified example, suppose the proposal first describes the product with the adjectives in the column below Version A, and then later uses the adjectives in the column under Version B to describe the same product:

Feature	Version A	Version B
color	blue	green*
size	small*	large
shape	boxy	sleek*
speed	fast*	slow

The customer wants the "features" marked with the asterisks. He accepts the Proposal with the understanding that the product will conform with either Version A or Version B, either of which contains enough desirable features to be satisfactory. However, the designer selects the cheaper feature from each **row**, resulting in a product with the underlined features. The customer is disappointed, since he got only one of his four preferences. In the arguments following delivery, the vendor repeatedly explains why Version A governs for color, while Version B governs for size. The customer remains dissatisfied.

After being burned in this way a few times, the customer learns to look for duplication in the specifications and proposals he receives. When he finds duplication, he says, "It's the old multiple-specification swindle again!", even though the vendor intended no deception. Professional proposal evaluators require vendors to respond once and only once to each specification requirement; they regard duplicate responses as a sufficient reason to disqualify the entire proposal.

6.3 Bullets (Not)

Bullets are words or short phrases, appropriate for outlines. They don't convey a complete thought, so they shouldn't be used in an FS. Bullets are very easy to misunderstand. Find an outline in bullet form of a talk you heard a year ago; try to remember the entire concept behind each bullet. Specifications use full English sentences, which evolved to enable the language to express complete thoughts.

6.4 Selective Emphasis (Not)

It's tempting to highlight a portion of the FS text, to emphasize the most important features or the result of a controversial decision. Highlighting may be done in many ways: bolding, italicizing, or underlining the text; using larger type font; using all upper case; using differently colored text; placing the text on a line by itself; drawing a box around the text; or simply by repetition.

The reader of the highlighted FS understands very clearly that the emphasized material sets forth important requirements that absolutely must be achieved. But what is the status of the remaining, un-highlighted text? Since the authors of the FS had the opportunity to highlight it but didn't, its importance is understood to be less than that of the critical highlighted requirements. The un-highlighted material might be interpreted as describing functions that aren't absolutely required, but optional features that are desired if they are cheap or convenient to implement.

This understanding is incorrect. The product must comply with all formally stated requirements throughout the FS, even those in its smallest footnote. Emphasizing certain first-class functions has induced second-class status onto all the other functions, raising doubt about their necessity. For this reason, selective emphasis should be avoided. If a requirement for a feature is stated once, that feature is 100 per cent required. Repeating the requirement for that feature on another page of the FS cannot increase that percentage, and leads to the problems discussed in Section 6.2. It may be desirable to separate requirements from other text, such as by numbering them, but all requirements should be treated uniformly.

6.5 Untestable Requirements

The writer of the Acceptance Test Procedure must translate each FS requirement into a test item. This process is frustrated by FS sentences that aren't statements of material fact about product functions, and imprecise requirements. For example, "X shall be approximately 3", doesn't allow the tester to determine whether the product complies. Is 3.1 close enough? How about 4? 17?

6.6 Contradictions

Internal contradictions are the most serious flaw in an FS. In a logical system, **any** contradiction completely destroys the system's ability to distinguish truth from falsehood; all statements are simultaneously true and false. Particular care must be taken to prevent conflicts between obscure sub-sections of standards that are referenced and the FS or other referenced standards. This is a sufficient motive to adopt a policy of referencing only very small and specific sub-sections of standards.

6.7 Impossible Precision

The FS will be interpreted literally. It must not require absolute precision or other impossibilities. For example, continuous variables, like the height of a device, should be stated as a range, or a value plus or minus a specified tolerance, rather than as a single number: "The height shall be 12 inches". Another absolute is that some output is to be produced "immediately". Signals cannot travel faster than light, and some processing delay is probably needed. When prompt response is required, a maximum time interval should be specified.

Another form of absolutism is the statement that something is to be done "to the greatest possible extent". This requires infinite resources, and is untestable.

6.8 UNNECESSARY REQUIREMENTS

An FS should be sparse, containing no redundant or unnecessary items. The Acceptance Test Procedure will include test items to verify all stated requirements, since the writer of the ATP isn't authorized to discard properly stated requirements that she believes are unnecessary. A seemingly simple requirement that is carelessly put into the FS can lead to a substantial waste of test resources, as well as design time and parts cost.

An example of a common unnecessary requirement is the inclusion of an entire Standard in the FS, when only a small portion would suffice. Another example is near-duplicate or redundant descriptions of a given function. The ATP author may not realize that only one actual function is being described, and develop several equivalent tests.

6.9 OVER-SPECIFICATION

While an FS must fully detail the product's functions, it shouldn't include extremely explicit descriptions. For example, if the FS includes a photograph of the front panel of a prototype, the designers aren't able to deviate from the initial panel in any way without departing from the FS. They can't change the color, the nomenclature, the positions of the controls and indicators, or even substitute a different model of knob or switch. The FS should describe all of these aspects of the front panel in detail, but shouldn't tie the designers' hands. The rule stated earlier applies: if a characteristic could be of importance to a customer, it should be covered in the FS, but not otherwise.

To see why exact compliance with the FS may be critical, suppose that our company has entered into a contract to deliver a quantity of turnkey computer systems in accordance with the FS. Suppose further that the FS requires our product to contain a specific computer model, the HAL 900. After spending several months customizing our software for this customer's needs, we try to order the HAL 900s to deliver as components of our product. The supplier informs us that during that time, their Model 900 has been replaced by the HAL 950. Model 950 is markedly superior to the old 900 in every way and costs $10,000 less. Of course, we are strongly motivated to make this substitution.

The customer wants the new 950, but he feels he should benefit from the price reduction. He says, "I had my heart set on the old reliable Model 900. I heard they haven't got all of the bugs out of the 950 yet." In one scenario, he offers to reluctantly accept the 950 if we reduce the price of each system by $15,000, since our cost plus overhead plus markup was lowered by this amount. In another case, he will accept the substitution if we throw in an uninterruptable power supply that he needs but forgot to include in the contract. In either case, we end up passing on the decreased cost to our customer, just because our FS called out a specific model number. If it had (correctly) specified the functions of the overall system (transactions per hour, latency) instead, we would have been able to substitute the Model 950, bringing the entire cost difference to our bottom line.

The moral is clear. By specifying unnecessarily precise details of our product, we needlessly give ammunition to a customer, should it turn out that we want to depart from the exact version we promised. If the Model 900 had become unavailable, we wouldn't have been able to fulfill the letter of our contract based on the FS at all. The customer could use our predicament to break the contract or demand a price reduction.

6.10 Proprietary Information

If the FS contains certain kinds of proprietary information, then it shouldn't be disclosed to non-employees. For example, the company generally doesn't want its customers to know its costs or schedules. If a customer knows that a particular product carries a high margin, he may be motivated to bargain for a lower price. If the actual development schedule were known, a salesperson would lose the option of promising unrealistically early delivery. A nondisclosure agreement is useless here; the threat isn't that the customer might give the information to the company's competitors, but that the customer himself will use the information in a way contrary to the company's interests. The inclusion of such proprietary material restricts the uses to which the FS may be put, limiting its benefits throughout the life of the product.

6.11 Open Issues

Before the FS is approved, all functional decisions should be made and the results recorded as requirements in the FS. There should be no remaining open issues, that is, topics to be studied and resolved in the future. This is a restatement of the principle expressed in Sections 5.2 and 8.5.2 that the FS must be complete. This is such an important criterion that it deserves repeating here. A large part of the power of a formal program derives from the principle that all functional decisions are made before implementation begins. Executives must have the determination to keep the product in its Definition Phase until the primary goal of that phase has been achieved.

7
OTHER DEVELOPMENT DOCUMENTS

The FS is the most important document that is used in formal methods, since it is the conduit of information between the two development phases. Its properties and content are presented in the last three chapters. To be successful, a formal method requires documents in addition to the FS. This chapter describes a number of documents, some of which should be required formal documents, while others are helpful optional or informal documents. Except where combinations of documents are suggested, all of these should be separate documents.

7.1 PRODUCT DEVELOPMENT PROCEDURE

Any formal program is based upon a written **Product Development Procedure** (PDP) that is one of the company's Quality Procedures. There is a single PDP that applies to the development of all products, whereas each product has its own version of each of the other documents described in the following sections. The PDP defines the organizational structure to be used for product development, the sequence of development phases, the set of mandatory documents, the optional documents, the review process, and the document control system. It defines the contents, structure, generation, approval, control, modification, and utilization of these documents.

The PDP should be written carefully to specify the complete development method. The PDP authors should assume that if a given activity isn't required, it won't be performed; if the activity is described as "optional", "elective", "conditional", or "discretionary", it won't ever be done; if an activity is required but defined vaguely, it will be performed at the absolute minimum level of effort, and probably fail to achieve the desired result. The PDP isn't a reminder to people who understand and accept formality; it is instructions to people with partial understanding of formal methods and limited inclination to cooperate. Within that framework, the following PDP ingredients are proposed.

7.1.1 Applicability, authorization, responsibility, and flexibility

The PDP must state that it applies to **all** of the company's future development activities, including new products and modifications, and to **all** employees, contractors, and subcontractors. It may be necessary to exempt existing products that have been developed informally, as well as products derived from them, and products presently undergoing informal development. Such products must be identified by name and specifically exempted in the PDP.

The individuals who are responsible for administering the product development process must be identified by title, not by name. The techniques they are to use to ensure compliance may be stated.

The individuals who are authorized to modify or waive any part of the PDP for a single development project must be identified by title, not by name. The degree of modification and the circumstances under which it is allowed to occur must be carefully limited. The objectives are to permit sufficient flexibility to allow each project to be tailored to accommodate its unique features [NASA1, pp. 11-14], without any possibility that the integrity of the formal development method may be compromised by taking shortcuts during some future "emergency". The instances in which it is most important to follow formal methods are exactly those instances in which it seems most urgent to revert to informally inspired methods. The PDP must be worded so as to absolutely prohibit such backsliding by **top management, executives, managers**, and lower level employees. A single poor executive decision can lose the benefits from many person-years of effort.

The Product Development Procedure should emphasize that executives have only the authority stated in the Procedure; unless specific authority is granted to the holder of a given title, that person doesn't have that authority. One necessary power that is frequently overlooked is the authority to **stop** a development project at any time. If some unforeseen shift takes place in the market, a product may become obsolete before its development is complete. The only sensible course may be to stop this development project immediately and re-deploy its resources toward a viable product. However, the PDP may provide for project termination only at Milestone Reviews. The possible actions are: (1) violate the PDP by stopping the project now, (2) amend the PDP retroactively to allow stopping now, or (3) continue spending resources until the present phase is complete, then stop the project. To avoid forcing a choice among these unattractive alternatives, the PDP should define a process whereby a project may be stopped, or suspended and resumed, at any time.

Most companies conduct development projects over a wide range of magnitudes, from the addition of a well-understood feature to an existing product to the creation of a new major product. The PDP should define a development process appropriate for the project of largest scope, and specify the modifications to be used by projects of lesser scope. One common modification is to allow certain specified documents to be combined (not eliminated) for small projects. For example, Architecture, High-Level Design, and Detailed Design Specifications may be combined for very small projects; see Section 7.1.4. Of course, WHAT and HOW material must never be combined into the same document.

Some quantitative rule must be supplied in the PDP to relate the level of required documentation to the scope of the project, to prevent all projects from using the documentation that is appropriate only for small projects. Warning: the formulation of this rule has proven difficult in practice; careful attention is needed here.

Suppose the rule is of the form, "Use full documentation when the scope of the project exceeds D dollars (or P person-hours, or M months), and reduced documentation otherwise". Since we want to establish at the beginning of the project which kind of documents are to be used, we might define "scope of the project" to mean the initial estimate of the project's size. If the size should increase beyond this estimate as the work is performed, we may end up documenting a large project with documentation intended only for small projects. Also, basing this decision on the initial estimate motivates some managers to deliberately under-estimate the size of the project, to "escape" the use of adequate documentation; they would rather over-run their estimate than learn to use appropriate methods. Their supervisors might tolerate this deliberate circumvention of the company's Quality Procedures in order to "save" the time and expense of the "extra" documentation.

Now suppose "scope of the project" is taken to mean the current estimate of the total project size; that is, the sum of the work completed plus the current estimate of the work remaining. A project could be started using "small" documents, then grow until its scope exceeds the threshold. At that time, the PDP would require the existing "small" documentation to be converted to "large" documentation, even though some of the "small" documents may have been completed and released. Note that the need to rework the documents occurs at exactly the time when an expansion of scope is detected, just when the pressure is greatest to expedite the remaining tasks. Although unpleasant, this alternative is preferable to accepting inadequate documentation or rewarding deliberate evasion of the Procedures. Perhaps a superior rule can be formulated. Section 7.1.4 contains some suggestions for combining documents.

The company-wide Quality Procedures should specify the process for amending the PDP and other quality documents. If not, the PDP needs to state how it may be amended and identify the individuals who are authorized to do this.

7.1.2 Define a formal method

The PDP must specify an **entire formal method**. The authors should not assume that the reader of the PDP has any understanding of such methods. The PDP isn't a tutorial, but must contain a complete definition of all processes to be followed.

A common error is to limit the PDP to a description of the documents that are to be generated. This shows very clearly that the authors believe that

Informal Method + Documents = Formal Method,

whose falsity is discussed as Bower's Inequality in Section 3.7 above. The employees will continue using their informal methods, since the PDP hasn't told them to do otherwise.

To ensure that this misunderstanding isn't reinforced by the PDP, the **use** of each document, its purpose, and **relationships** to other documents must be included, along with clear statements of fundamental formal principles. The PDP must be reviewed carefully, using a literal interpretation of its wording, to verify that it is complete.

7.1.3 Definition of phases

The PDP may define a number of development sub-phases. The sub-phases must be given names and defined so as to be disjoint (not overlapping) in time. The **sequence** of the sub-phases must be defined. The FS must be completed, reviewed, approved, and released **before** the phases containing architecture and design **begin**. Thus, at least two phases are required; more sub-phases may be added to provide management checkpoints as desired, as discussed in detail in Chapter 9 below. The development activities in each phase should be allocated to departments. The deliverables (documents and other tangible items) for each phase must be defined. If some strategy other than a linear "waterfall" is desired, its phases must be carefully defined, and precautions should be built in to prevent the strategy from degenerating into an everything-at-once informal method.

7.1.4 Definition of documents

The PDP must specify the names and contents of the required documents. A description of the required contents is needed in the PDP to ensure that all of the necessary issues are actually addressed; an outline of the document is inadequate. If additional documentation is desired for a given project, it should be allowed but not defined by the PDP.

All required documents must be formally reviewed, approved, and released to the company's document control system. The individuals who must approve each document type must be identified by title. The PDP must require all approvers to declare their acceptance of the document by signing it, either on its first page or on a supplementary signoff sheet.

The PDP should establish nomenclature for the required documents. The names that the PDP mandates for each document aren't important, but these names should be required to be used consistently across all company products. For example, if the FS is required to be named the Functional Specification, then the PDP should require that the title of the FS must contain the character string, "Functional Specification", and prohibit all other documents' titles from containing this string. This requirement helps to make it clear that there is exactly one FS, and identifies it with certainty.

If such a nomenclature isn't mandated, an author might name his document the "Product Spec". When Quality Assurance seeks the FS, the author says, "This is the document you're looking for. We decided to name it Product Spec, instead of Functional Specification, to emphasize that a **product** is being described, etc. Go ahead and check off the FS on your document list. It's complete." When someone else complains because this document doesn't contain some of the material required of an FS, the author then says, "As you can plainly see, this document is just the Product Spec, not the whole FS. I think Mary is working on the document you're looking for." Mandating exact nomenclature prevents duplicity of this sort.

The formal documents that are to be generated and released during the development must be listed by title in the Development Plan (see Section 7.6). Thus, the Development Plan (as well as the FS) are required documents. The PDP should offer a choice of documentation scopes to match the complexity of the development project (see Section 7.1.1). One strategy is to combine various design specifications for projects of lesser scope, thereby allowing fewer review cycles without omitting any essential information. Large projects have a separate Architecture Specification (AS; see Section 7.3), High-Level Design Specification (HLDS; see Section 7.4), and Detailed Design Specification (DDS; see Section 7.5). In fact, very large projects may partition their product into modules (sub-assemblies) and generate an individual HLDS and/or DDS for each module. Or a separate document may be written for each major technology (hardware, software, mechanical, etc.) that is employed. For medium-sized projects, it may be appropriate to generate an AS, then combine the HLDS and DDS into a single document. For small projects and post-design modifications, the AS, HLDS, and DDS can all be combined into a single Design Specification. (Of course, the FS cannot ever be combined with design information.)

The PDP must require all of the formal documents to be kept current after they are released. When a change occurs, all documents within the scope of the change are revised, reviewed, approved, and released as a new version number. (For small changes, all this work can be done at a single meeting.) For example, if a module were split into two modules, the HLDS and the corresponding parts of the DDS would be changed, but the FS and AS would not.

7.1.5 Description of reviews

The PDP should define the process whereby formal documents and other deliverables are to be reviewed, as presented in Chapter 8 below and in [YOURDON, pp. 153-175]. The titles of the individuals who must participate in the reviews are stated; this list generally differs from the list of approvers discussed in Section 7.1.4 above. The activities during the review meeting and the two possible outcomes of the review should be defined in the PDP, as set forth in Section 8.3. The PDP should require the Scribe to write and distribute the Minutes of the meeting, as set forth in Section 8.4. The PDP must state that the review process must be repeated until a result of PASS is obtained, followed by release of the deliverable to the document control system.

7.1.6 The Document Control System

The PDP should define a document control system. This system must establish a method for numbering successive releases of documents, and allow an employee to learn the current (most recently approved) version number. Persons who are authorized by management to access documents should be able to obtain a read-only copy of the latest version, while access should be denied to unauthorized persons. After the MR review process (see Section 4.8.5 above) has authorized a specific modification to a document, the author or maintainer must be allowed read-write access to make these changes. The revised document is assigned a different revision number. The new revision must be approved by the signatures of people with the same titles that signed the original release of the document, and then released into the document control system. This system must prevent changes attempted in any other way.

7.1.7 The Engineering Process Group

The PDP should establish an Engineering Process Group, which is a team of definers and designers who are charged with monitoring the effectiveness of the current application of the PDP and formulating continuous improvements in this document. Working engineers spend part of their time participating in this Group. The Engineering Process Group should coordinate its efforts with the Quality Assurance Department. The results reported in Section 12.3.5 below indicate that improvement can be achieved at a satisfactory rate if one per cent of the Engineering Department budget is spent for salaries of people while they work in the Engineering Process Group.

7.2 WISH LIST

The Wish List is an informal document that is written by the Marketing Department early in the Definition Phase. The description of this document in Section 4.6.1 won't be repeated here. A highly informal name was deliberately assigned to this document to emphasize that it isn't a rival to the FS in any way. See Section 4.9.3.

7.3 ARCHITECTURE SPECIFICATION

The **Architecture Specification** (AS) is a mandatory formal document, written at the beginning of the Design Phase. Its reference is the FS.

The first job of the AS is to explain the partitioning of the product into modules and sub-modules that implement portions of the whole product. A Block Diagram is very helpful to show the relationship among these modules and to label their interfaces. This Block Diagram differs from the functional block diagram in the FS (see Section 5.1) that shows external interfaces, and whose functional blocks may differ from the modules that arise from the architecture. Modular partitioning is usually done according to technology (such as digital versus analog, or hardware versus software), and/or by physical assembly, such as a number of printed circuit boards. For complex structures, the top level Block Diagram may be supplemented by block diagrams of each of its principal modules, showing their sub-modules and interfaces in turn. A narrative explanation must accompany each block diagram, organized to show the flow from the diagram's inputs to its outputs, and explaining the functions of each sub-module. Since the AS is a high-level document, partitioning to lower-level sub-sub-modules should be avoided. The optimum number of layers to include in the AS depends on the complexity of the product.

The AS must assign names to these modules and sub-modules, and the internal interfaces among them. (The external interfaces have been given nomenclature by the FS.) The AS (and all other documents) must use the FS nomenclature and its own naming conventions consistently.

After the modules and sub-modules have been identified, named, and explained, the AS allocates the functions from the FS among these modules and sub-modules. Detailed descriptions of each module and its sub-modules are provided, including the numbers of the FS sections and sub-sections whose functions are implemented by the module being described. All sections and sub-sections of the FS that place functional requirements on the product must be mentioned in the AS to provide functional traceability back to the FS. The review of the AS will verify this relationship. The AS must use a hierarchical section numbering scheme, so that the other design specifications can reference specific sections and sub-sections of the AS in turn.

It is increasingly important to leverage the work of the designers by implementing a given function once, and using the design in several products. Reusing a module from a previous product that was designed without planning for reuse has proven difficult. To be fully effective, modules must be originally designed with possible reuse in mind. The AS should address the topic of reuse in two ways. First, the AS should identify the existing modules and assemblies that can be reused to help implement the current product, including any known top-level modifications that will be needed. Second, the AS should include a plan for implementing the current product with modules that can also be used to implement future products. The formulation of these plans depends on the technology in use, and is beyond the scope of this book. See [NASA1, pp. 15-17].

The AS should address issues that affect products throughout a product line, or group of related products. For example, it may be desirable to design the product so that its internal interfaces are compatible with certain company standards. Since these interfaces aren't observable externally, they aren't functions, and cannot be specified in the FS. Thus, the AS is the proper vehicle to use to require the implementation to use these interface standards. Such architectural requirements cannot conflict with the functional requirements in the FS, and should not be more detailed than necessary, to avoid tying the designers' hands.

Often a desire for product flexibility or future expansion is manifested as a request in the FS that the designers "consider" these topics. As discussed in Section 6.1.4 above, a request that the designers give consideration to some topic isn't a function of the product, and should not be included in the FS. The AS is the proper place for this material. The architects can communicate this architectural requirement to the designers by including a requirement in the AS for a specific quantity of resources to be dedicated to modification or expansion. For example, "Each circuit board shall reserve at least 12 per cent of its useful area for expansion". (It is assumed that the existence of "circuit boards" has been established in the partitioning section of the AS.) To prevent the designer from evading the spirit of this requirement by counting all of the useless scraps of area between devices as parts of the expansion area, the further requirement, "The expansion area shall be marked with a convex boundary", may be included. The strength of this approach is that it provides a testable criterion to ensure that the designers will meet a somewhat intangible requirement.

The AS is an internal document, not generally disclosed outside the company. Therefore, the AS may contain proprietary material. In particular, the AS should define any special algorithms that are to be used in the design. A general specification of the algorithm, such as a mathematical formula, should be used instead of a specific implementation. The designers must be free to use any implementation that yields a functionally equivalent result.

The AS should identify the foreseeable technical risks that might prevent the functions in the FS from being achieved, or might make the development longer or more costly than anticipated. The AS should require the High-Level Design Specification to contain a plan for addressing these risks by estimating their probabilities and their consequences, and proposing remedies for recovery if the risks should materialize.

If the implementation involves a processor, the AS should identify all events that can cause interrupts and specify their priority and required latencies.

When it has been released, the AS will serve many purposes. Before it can start doing any of these things, it must first be reviewed and approved. Thus, its first goal is to pass its own review. To do so, it must **convince its reviewers** that complete functional coverage has been achieved, that a competent foundation for implementation has been provided, and that no superior alternatives exist (see Section 8.6). Therefore, one of the roles that the AS plays is that of persuasion. Since no design yet exists, the AS shouldn't be expected to show that the functions in the FS will be met; all the AS can do is to provide a framework in which a competent design can be expected to meet all FS requirements.

7.4 High-Level Design Specification

The **High-Level Design Specification** (HLDS) is a mandatory formal document, written during the Design Phase, after the Architecture Specification has been released. Its references are the FS and the AS, so its nomenclature and concepts must be consistent with those two references.

The HLDS completes the partitioning that was started in the AS. It partitions the modules and sub-modules from the AS into as many layers as are needed to reach the lowest-level sub-...-sub-module that is the "atomic unit" of implementation. For example, a C-language software module is partitioned into individual functions. As in the AS, block diagrams are helpful to show the relationships among units. Names are assigned to all modules, as well as to all elements of each interface between units. Each such interface must be functionally specified [NASA1, p. 8]; if it is a digital signal, for example, the signal names, voltage levels, polarity, coding, waveforms, frequency, and protocol (if appropriate) must be detailed. To determine the required level of detail, keep in mind that individual units may be designed by different (groups of) designers; if both designs meet the interface definition, the units should interoperate.

All sections and sub-sections of the FS that place functional requirements on the product must be mentioned in the HLDS to provide functional traceability back to the FS. The review of the HLDS will verify this relationship. The HLDS should also reference key sections of the AS by number. The HLDS must use a hierarchical section numbering scheme, so that the other design specifications can reference specific sections and sub-sections of the HLDS in turn.

The HLDS should contain the risk analysis required by the AS. If the product stores data, the internal data structures that are shared by two or more of the lowest-level design units should be defined by the HLDS, including organization, access method, content, and meaning of each valid data value. If the FS specifies any time limits for the product to act or respond to an input, these real-time issues should be analyzed in the HLDS. This document should convince its reviewers that a design based on the HLDS will meet all FS requirements.

For complex products that use multiple technologies, the HLDS may be broken into parts; for example, a Hardware HLDS and a Software HLDS, since these parts may be written by different authors. It is important to review all of the parts together, to detect differences in interpretation among the various design groups.

7.5 DETAILED DESIGN SPECIFICATION

The **Detailed Design Specification** (DDS) is a mandatory formal document, written during the Design Phase, after the High-Level Design Specification has been released. Its references are the FS, the AS, and the HLDS. Since it is driven by these documents, the DDS must use their concepts and nomenclature consistently, and must describe a detailed implementation that complies with these references. The DDS refines and extends the approaches selected in the AS and HLDS [NASA1, p. 8]. The DDS must describe the elements necessary to implement all of the FS functions and no others.

Usually each atomic sub-...-sub-module (unit) has its own sub-section in the DDS, giving the name of the unit, references to the FS and HDLS sections being implemented, a brief functional description, an interface description, and the detailed design of the unit. If the unit were electronic hardware, a low-level block diagram or schematic could be used to express the detailed design; for software, a flow chart or (preferably) pseudocode could be used. Since the DDS is written and reviewed **before** the detailed design is completed, a detailed schematic or source code is not yet available (or desired).

7.6 Development Plan

The **Development Plan** is a mandatory formal document, typically written concurrently with the AS at the start of the Design Phase.

The Development Plan contains the information needed to administer the development project, including personnel assignment, detailed milestone schedule, budget, development tools to be used, capital equipment and the cost of equipment to be purchased, and a list of realistic assumptions that form the basis of this material. An example of a realistic assumption is to allow time for two review meetings for each document.

The Development Plan must document all administrative choices that affect this project. Decisions must be made for each optional aspect of the project: whether to treat this as a "small" product and combine certain design documents, which optional processes are being waived, and how this development is tailored to the specifics of the product. All such decisions must be made and recorded in the Development Plan; no open issues are allowed.

7.7 Acceptance Test Procedure and Test Report

The **Acceptance Test Procedure** (ATP) is a mandatory formal document, written during the Design Phase, typically by Test Engineers. (Some organizations refer to this document as a Test Plan, a First Article Test, a Proof-Of-Design Test, or an Acceptance Test.) Its purpose is to define a sequence of detailed tests (called test items) to be applied to one (or a few) pre-production units to validate the design; that is, to demonstrate that the implementation correctly performs all of the FS functions. It isn't a Production Test, to be applied to all units as they are manufactured.

The ATP is driven by the FS and nothing else. The author(s) of the ATP and the designers must independently follow the same standard procedure to determine what the product does, to ensure that the design passes the acceptance test. The standard rule is: **Every statement of material fact in the FS that defines a function of the product leads to a test item in the ATP.** Each test item references the section number in the FS whose function(s) it tests. The ATP can be augmented to also confirm that the architectural requirements stated in the AS are being met. Whereas the product is tested as a "black box" to validate its functional requirements, it is considered as a "white box" to validate its architectural requirements; the "interior" of the product must be examined. To emphasize this difference of viewpoint, it is recommended that these architectural tests be segregated into an Appendix to the ATP.

The ATP should be organized as a sequence of test items, including the product's mode of operation, environment, initialization, inputs, and expected outputs. If an output can have more than one correct value, the acceptable tolerance or range of values should be specified, so that the tester can definitely determine whether or not a test item has passed. The results of all tests should be recorded in a Test Report.

Since the ATP is based only on the FS, its authors can begin writing it when the Design Phase starts and finish without waiting for any other document. The architectural Appendix to the ATP (if any) can be finalized as soon as the AS is released. The ATP should be completed, reviewed, approved, and released before the design is complete, so that testing can begin without delay.

7.8 Modification Request

The Modification Request is described in Section 4.8.4. The company probably already uses MRs to track changes during production, so this isn't a new document type.

7.9 Future Feature List

The Future Feature List is an informal document that could be maintained by a marketer or a systems engineer. It collects ideas for features that aren't included in the version presently being developed, but that might be included in future versions of the product. Ideas can be added informally by anyone at any time. When a feature is nominated by a Modification Request but isn't accepted by the MR reviewers for inclusion in the present version, it should be recorded in the Future Feature List.

7.10 Meeting Minutes

The Minutes of the review meeting should include all decisions made, all agreed changes, **all open issues**, and a clear statement of whether the outcome was PASS or FAIL. The Minutes should be organized by topic, not as a chronological "he said, she said" narrative.

8

THE REVIEW PROCESS

The review process is an important element of a generic formal development program. A review consists of preparation, a review meeting, and follow-up activities. All formal documents and other deliverables must pass their reviews.

8.1 PREPARATION FOR A REVIEW

The material to be reviewed must be distributed to all reviewers some minimum length of time before the review meeting; three business days is about right, while five is too many. (The author shouldn't change the document between distribution and review, so she is "idle" during this interval.) The reviewers must prepare thoroughly before the meeting. Suggestions for preparing to review an FS are given in Section 8.5, while specific preparation for reviewing a Design Specification is in Section 8.6. The reviewers can decide who will play which role during FS preparation; each reviewer doesn't have to play each role. Reviewers are encouraged to submit written comments to the author(s) before the meeting, particularly if substantial problems are discovered. The amount of time spent in preparation should exceed the time spent during the review meeting.

If some of the reviewers work at remote locations, extra time might be allowed for them to travel to the review site. An attractive alternative is to allow them to participate via videoconference. Reviews over audio teleconferences have proven disappointing. Remote reviewers may be allowed to submit written comments in advance as a substitute for attendance.

8.2 THE PURPOSE OF A REVIEW

The purpose of a review is to verify that the deliverable(s) being reviewed are of sufficient quality to serve as permanent records of the work they describe and to guide the following tasks. Upon successful completion of the reviews of all deliverables of a development phase, management has the option of terminating or suspending the project, or authorizing work to begin on the following phase.

It may seem efficient to complete all of the design tasks without intermediate reviews, then conduct a single overall review to catch all faults. However, different questions are asked and answered at the reviews that end each phase, so that high-level defects can be found and repaired before the high-level deliverables are used to drive the lower-level work. If these reviews are skipped, a large quantity of low-level work may need to be re-done to perform each high-level change. Experience [YOURDON, p. 27] reveals that it isn't acceptable to skip intermediate reviews.

8.3 CONDUCTING A REVIEW MEETING

The activities during the review meeting should follow a common pattern. At the start of the meeting, reviewers are appointed to be Presenter, Scribe, and Moderator. The Presenter (usually an author) first makes an overall conceptual presentation of the document, followed by a section-by-section detailed presentation. The Scribe takes notes, making a careful record of all changes to the document that the reviewers agree on. (If consensus isn't possible, the PDP must specify a decision-making process.) The Moderator keeps the meeting on topic and moving forward. These three tasks should be done by separate people, if there are enough reviewers.

The reviewers should try to **identify all** of the defects and open issues in the deliverable being reviewed, so as to minimize the number of iterations of the review. The reviewers should **not** try to **resolve** the problems they have identified, so as to allow the entire deliverable to be reviewed within a reasonable time; a limit of two hours per session is recommended. The PDP should provide for adjournment and resumption of a review meeting.

A review consists of evaluating a deliverable against a **reference**. The FS is the first formal document to be reviewed, so no formal reference is available for its review. In this special case, the informal Wish List may be consulted, but the combined experience of the reviewers is the ultimate authority. Since no formal reference is available for reviewing the FS, the specific techniques detailed in Section 8.5 have evolved as a substitute. The reference(s) for the other formal document are defined in Sections 7.3 through 7.7. The references should be viewed as "cast in concrete"; the reviewers must focus entirely on the deliverables under review. The review meeting is emphatically **not** an opportunity to discuss changes to the references.

The focus of the review should be on checking compliance with the references. Any non-compliant approach or design must be rejected, no matter how attractive it may be on other grounds. Once full agreement with the references has been established, questions can be raised concerning whether superior alternatives exist.

In the absence of a competent reference, a deliverable can be criticized or admired, but it cannot be formally reviewed. This is a major stumbling point for informal developments that imitate formality by generating some documents and attempting to review them. If a key document has been omitted or if the development tasks haven't been performed in the proper sequence, the references are absent or incomplete, so a comparison between the references and the deliverable under review isn't possible. Without the benefit of a valid review, the document being reviewed won't be available as a robust reference for reviewing the next document. Once the chain of strong documents is broken, formality cannot be reinstated at a later phase for this reason.

The PDP should define the two possible **outcomes** of the review:

> PASS: The defects in the deliverable that were discovered during the meeting are minor. The author is directed to make the changes that were agreed upon and promptly submit the modified deliverable for approval. No further meeting is needed.
>
> FAIL: Significant defects were identified at the meeting. The author is directed to amend the deliverable in accordance with the agreed changes, re-distribute the result, and schedule another review meeting.

At the end of the review meeting, a decision must be reached as to whether the outcome was PASS or FAIL; there are no other alternatives. The PDP should specify how this decision is to be made if consensus cannot be obtained.

8.4 After the Review

The Scribe should write the Minutes of the meeting, as described in Section 7.10. These Minutes are required to be distributed to the reviewers within some fixed number of business days (such as one or two). The author(s) of the deliverable should make the agreed changes, investigate the open issues, make decisions, and record those decisions in the deliverable. If the outcome of the review was PASS, the deliverable is circulated for approval signatures and released to document control. If the outcome was FAIL, another review meeting is scheduled and the revised deliverable is distributed to the reviewers.

8.5 SPECIFIC TECHNIQUES TO REVIEW AN FS

Let's suppose that our team of authors has completed the FS for a new product and submitted it to the FS reviewers for approval. How should these reviewers evaluate the FS, to see if it is worthy of management's signatures? That is an important question, since many person-hours of effort will be based on the wording of the FS, and later changes will be very expensive. Clearly, we need to give the proposed FS a thorough review, but how?

We can start by looking for the desirable and undesirable items listed in Chapters 5 and 6. After this mechanical process, the FS is subjected to an extensive analysis. The reviewers examine the FS from the points of view of all classes of users of the FS. The reviewers **play the role** of each class, one at a time, to evaluate the suitability of the FS for the purposes to which it will be put by that class. (Some other specific FS review techniques are given in [YOURDON, pp. 153-163]). Management can ask an employee from another project or division to participate as a reviewer, or they can hire a consultant for this purpose. It might be possible to find someone who recently retired from a customer, or a trusted employee of a friendly customer to help. While such assistants may give valuable advice, management's personal participation in the review process is essential.

Reviewers should seek out harsh "environments" for the FS as they conduct their role-playing thought experiments. This is done by analogy with the "burn-in" process, in which hardware devices are subjected to abnormally stringent conditions to screen out weak items. Even though we don't expect that the FS will ever be used for a particular purpose, a worst-case test can ensure that the FS is ready to meet an unforeseen requirement. For example, a salesperson may give in to the demands of an important customer for functional details by "losing" a copy of the FS in the customer's in-basket, contrary to management's wishes for confidentiality. Careful construction can produce a single FS that meets all of the diverse purposes that an FS can serve. By assuming that the readers of the FS work for other employers, aren't under the reviewers' direct control, and have a conflict of interest with them, reviewers can ascertain to what degree the FS protects the firm's interests. In this way, many flaws in the FS can be revealed and repaired before actual harm is done.

8.5.1 Reviewer as User of the product

During the generation of the FS, the authors and reviewers should have spent considerable energy **visualizing** the product, pretending to operate it, and imagining how real users would interact with it. If this is done adequately, the product seems real to the FS authors; it is right here, right now, just beyond the authors' fingertips. Surely the authors know the product's size and shape, know what it's connected to, know what's under it, and know what color it is. (And the authors automatically write about the product in the present tense, not the future tense.) If the FS authors don't know these elementary facts about the product, they haven't been visualizing the product; instead, they have just been manipulating text in a word processor. The value they have added is clerical, not intellectual.

8.5.2 Reviewer as Customer for the product

The reviewers play the role of a potential customer who has been given a copy of the FS. To establish "harsh" conditions for this thought experiment, the reviewers suppose that this reader has **no other source of functional information** about this product. The "customer" may ask questions about price, warranty policy, delivery schedule, and all other non-functional areas, but is prohibited from supplementing the functional information in the FS from any outside source. The purpose of this exercise is to see if the FS is complete; whether it supplies **all** of the technical information that the "customer" needs to make a purchase decision. For example, does the FS alone allow the "customer" to evaluate his employees' ability to install, configure, and operate the product? The reviewer notes the areas in which the FS is insufficient, so that these topics can be augmented before the FS is approved.

As the experienced reviewer reads the FS, he sees if it provides satisfactory answers to his mental list of stock questions, corresponding to areas that have often been overlooked in the specifications he has previously read. For instance, it is often informative to see if the FS answers the question, "What color is the product?" There are a few products (like gasoline) whose colors are of no importance to their users, but color is almost always of interest, and is frequently overlooked in specifications. (The way to determine whether color should be specified is to try to imagine a color, or combination of colors, that could make a customer react negatively. "Is that real vomit all over your product?")

The most difficult aspect of checking an FS for completeness isn't spotting missing details, but recognizing that some necessary major topic has been completely omitted. Mistakes are relatively easy to find because they are present in the document. Omissions are much harder to spot because…they're invisible! If an FS author is uncertain about whether a given topic belongs in the FS, she should write one sentence in the FS explaining why the topic isn't applicable. If all of the reviewers agree, no harm has been done. If this sentence causes any reviewer to think about whether the topic should be addressed, then a valuable service has been rendered, because the reviewer might not have done so otherwise.

8.5.3 Reviewer as Buyer of the product

To explore other aspects of the FS, the reviewers now pretend they are buyers of the product. The "buyers" pretend they have solicited specifications from a number of vendors, and have received the FS being evaluated along with specifications from competitors. Their objective is to judge a real buyer's reaction to the FS.

The Saga of the Refrigerators

Suppose the buyer wants to purchase 100 industrial refrigerators. He has received formal specifications from competing refrigerator vendors, and our informal FS that doesn't use formal specification language or other conventions. What does the buyer think when he reads our FS?

His first reaction is, "How can these guys have been in the refrigeration business as long as they claim without learning how to write a specification? Surely they have bought components according to specs, have seen their competitors' specs, have bid on customers' specs, and previously worked for companies that specified their products. They cannot have escaped noticing that all of these documents used the same formal language. Why in the world have they submitted an inferior document to me?"

His second thought is, "This thing must have been written by a new junior employee inexperienced with specs. That vendor has so little interest in my business that they didn't even give their document a cursory review."

His third reaction is, "That vendor knows very well how to specify its products, but they avoided stating any formal requirements in this document. They pretended to promise performance, without actually doing so. It's a deliberate attempt to defraud me! If I enter into a contract based on this document, they will unload a whole junkyard of obsolete, broken refrigerators onto my receiving dock and demand payment. When I object, they will correctly say, 'That document is a story about refrigeration principles, to illustrate our competence. As you can plainly see, it doesn't actually require our product to keep anything cold.' If I exchange my genuine promise to pay for their deceptive non-promise to deliver, I'll be in court for the next five years."

All of these impressions of our informal document are incredibly negative. The buyer doesn't care which possibility is true, since he has already discarded our FS, and is focused on our competitors' competent specifications.

8.5.4 Reviewer as Seller of the product

In this scenario, the reviewer play the role of a vendor, who contracts to deliver a product meeting the FS. (If this is the reviewer's real job, not much imagination is needed here.) The reviewer pretends that she is **really** entering into a binding contract to deliver a product according to the letter of the FS, and searches for any possible adverse consequences.

There are several aspects to consider. When the reviewer visualizes a customer reading the FS, she realizes that the FS shouldn't contain any "sensitive" information. The avoidance of cost estimates and development schedules has been discussed in Section 6.10.

The principal question to be addressed is whether the FS is too strict. Does it state requirements that are absolute, or that may be difficult to meet when interpreted literally? The example of a photograph of the product is discussed in Section 6.9.

The reviewer also explores the clarity, rigor, and precision of the FS by asking, "How can it be misunderstood?" The reviewer looks for ambiguous language. If the FS says that the product supplies "A and B or C", this could mean that either C must be present, or that both A and B must be present. It could also mean that A is always present, and that either B or C is present. Who decides which interpretation governs? The vendor and customer may each assume that they will be allowed to make this choice, and they are likely to have conflicting goals. To avoid disappointment and litigation, the FS must be cleansed of fuzzy expression, sloppy language, and ambiguities.

8.5.5 Reviewer as Buyer of contracted development

The reviewer imagines that he enters into a contract with an outside contractor, who agrees to design a product in compliance with the FS, and to deliver a quantity of production units, in exchange for a fixed fee. To generate a harsh environment, the reviewer pretends the contractor is a very experienced loophole-finder, that the reviewer's firm will have no opportunity to disapprove the product before production, that the contractor isn't dependent on the reviewer's firm for future business, and that both the production quantity and fee are large. (If extra incentive is needed, the reviewer imagines that all of the production units will be stored in the reviewer's own personal office until they are sold.)

In Section 8.5.4, the reviewer was the seller of the product, and was concerned about an FS that was too strict. The shoe is on the other foot now; the reviewer is buying the product, so the relevant question becomes, "Is the FS too lenient?" That is, can the contractor find some way to make a product that complies with the letter of the FS, but is still unsatisfactory? This isn't an unrealistic situation, since the contractor has strong financial motivation to reduce the cost of the product, and a less expensive component or production process may well yield an inferior product.

Suppose that the production lot is delivered, and we are shocked to see that some of the units are gray and others are various shades of muddy brown. We say, "We expected a uniform sky blue; didn't you see our mockup?" The contractor replies, "The FS says the exterior shall be painted. For each production run, we mix together the paint left over from our other jobs, because that's cheapest for us. The mockup isn't mentioned in the contract." At this moment, we realize that there is a deficiency in the FS; a feature of importance was overlooked.

Another common source of problems is the specification of a contractor's **process**, rather than the **result** to be achieved. The contractor carefully follows the required process to the letter, without regard for the results obtained. The results are unlikely to be satisfactory, but the contractor cannot be held accountable because the specified process was followed. Section 6.1.4 warned of this situation; this portion of the role-playing exercise reveals its presence.

The reviewer attempts to anticipate all such negative possibilities via role-playing, so that a lenient FS can be reinforced before it is approved. The reviewer applies the following test: Is the FS so robust that any reasonable interpretation results in an acceptable product? The reviewer assumes that he has no opportunity to review the design or make any "midcourse corrections"; instead, he must accept any product that complies with the letter of the FS. This is a realistic assumption, since the contractor will charge a very high price for functional changes after the contract is signed.

8.5.6 Reviewer as Development Contractor

To take the process a step further, the reviewer switches sides in the previous scenario, and now pretends to be the designer who has accepted the contract to develop the product. In this role, he aggressively seeks design alternatives that reduce his cost while complying with an interpretation of the FS that is as favorable to the contractor as possible. The reviewer keeps in mind that the designer is **trying** to misunderstand the FS, to reduce his cost. From the contractor's point of view, if an unsatisfactory product results, **so much the better**! He then has an opportunity to sell the required corrections to a captive buyer. Companies that develop products under government contract must become extremely proficient at this practice.

As a favorite example, consider an FS that says, "When Condition A exists, Indicator 1 shall be ON." An experienced contract designer will smile as he wires Indicator 1 across the power supply, keeping it ON all the time. He knows that the test item in the Acceptance Test Procedure **must** say, "Generate Condition A. Observe that Indicator 1 is ON.", since that is the literal FS requirement. He knows that his "design" will pass this test, since Indicator 1 is always ON. A proper FS prevents this "misunderstanding" by requiring that, "Indicator 1 shall be ON if and only if Condition A exists." This requires a much more extensive test procedure that verifies that Indicator 1 is OFF under all circumstances when Condition A is absent, but this is necessary to compel the designer to provide what the user wants.

Clearly, it is important to prepare a rigorous, high-quality formal FS when the product is to be developed by an independent contractor. Yet somehow we seem to think that it's acceptable to require our employees to develop products based on informal "specs". Look me in the eye and explain why our employees should be forced to work with tools that are inferior to those we provide to outsiders. Is it because we imagine that it will be possible to make low-cost changes to the "spec" after the design has started? An enormous quantity of experience has shown that this is a very unrealistic and expensive assumption.

8.5.7 Reviewer as Middleman (Dealers!)

In previous sections, the reviewer has acted as the seller of a product based on an FS, and then as the buyer of a product based on an FS. To unite these concepts, the reviewer can imagine that she performs both functions. Every morning she procures products from outside developers and manufacturers, and every afternoon she sells those products to customers, with all of these activities based on the **same** FS. If the reviewer is willing to buy and sell according to the FS, then no loopholes are visible for either party to exploit, and the FS may be submitted for approval and then used in every way presented in Chapter 12 to obtain the resulting benefits. This is the goal the reviewers seek.

8.6 SPECIFIC TECHNIQUES TO REVIEW DESIGN SPECIFICATIONS

For the purposes of this section, the three types of Design Specifications are the Architecture Specification (AS; see Section 7.3), the High-Level Design Specification (HLDS; see Section 7.4), and the Detailed Design Specification (DDS; see Section 7.5). Multiple documents of the same type may be reviewed at the same meeting for convenience, but documents of different types shouldn't be reviewed together, because their goals are different. (Of course, if two or more types are combined into a single document for a small project, then the whole document should be reviewed at a single meeting. Another exception is a minor change made after all documents have been released, in which case all documents should be reviewed together.)

As with all documents, the Design Specifications should be reviewed for completeness, clarity, top-down organization, consistency, and for meeting their specific goals as set forth in their respective sections of Chapter 7. In addition, these documents should be carefully tested to verify that they address all of the functional requirements set forth in the FS [YOURDON, pp. 164-175]. A subtle distinction is needed. When the AS or the HLDS is being reviewed, no detailed design has been synthesized yet, so the appropriate questions are: "Is the framework being reviewed consistent with the FS? Is it likely that the approaches outlined in the document under review will lead to a compliant design?" When the DDS is being reviewed, it contains a specific low-level design, so the question changes to: "Does the proposed design implement the FS?" This question cannot be answered by the two high-level documents alone.

This verification process can be mechanized by using a Compliance Matrix [NASA1, p. 26]. This matrix isn't a formal document, just a tool to help the reviewers. Each row of this matrix corresponds to a functional requirement; the columns correspond to the section and sub-section numbers of the document(s) being reviewed; the element in a given row and column of the matrix indicates whether its document section (column) states that its requirement (row) is being met. The rows are derived from the Table of Contents of the FS by deleting any FS section number that doesn't state any functional requirement. (This derivation needs to be done only once whenever the FS is amended; the set of requirement-bearing sections can be used to review all of the Design Specifications that are based on that FS version. This can be done on hard copy or in a file or database.)

The Compliance Matrix is filled out by searching the documents being reviewed for the FS section or sub-section number of each row, and marking each box where a match is found with an 'X'. The string search (Find) capability of a word processor can be used to identify candidate sections, which are then read to ensure that they promise actual compliance. One of the reasons to include FS section numbers in the design documents is to facilitate this process by making functions traceable back to the FS. After completion, the Compliance Matrix is analyzed to verify requirement coverage.

If there is a requirement (row) without an 'X', we have found a requirement that "slipped through the crack"; it isn't addressed by any section of any document under review. The requirement should be allocated somewhere, and the corresponding document should be amended. If a row contains a single 'X', the sections in the FS and document under review should be carefully read to ensure that all functional requirements in that FS section have been addressed. If a row contains two or more 'X's, this isn't necessarily an error; the functions may be implemented by a combination of the mechanisms described in these sections. All sections should be read to verify that the functions have all been addressed, and to resolve duplications.

9

DEVELOPMENT PHASES

Formal development has been explained in terms of two development phases, called the Definition Phase and the Design Phase. The reason for dividing the time line into two disjoint segments is to ensure that the FS is completed before the design begins. This is the central issue that separates informal and formal methods; it is a source of the great strength of the formal approach.

There are other formal tasks that also need to be performed sequentially. While doing the other tasks in order isn't as crucial as the FS-design sequence, it is nonetheless important. One way to ensure that the proper sequence is followed is to extend the phase mechanism to include the other tasks. The Design Phase is subdivided into additional phases; a convenient sequence of phases follows:

Definition Phase

Architecture/Planning Phase	
High-Level Design Phase	
Detailed Design Phase	———Design Phase
Implementation Phase	
Acceptance Test Phase	

Management uses these phases to control and monitor the progress of the development. The proper sequencing of tasks can be enforced by making the entrance criteria of each phase identical to the exit criteria of the previous phase [NASA1, p. 7]. These suggested phases are described in the following sections. These generic activities and documents may be supplemented with tasks and documents relevant to a specific industry and company to generate a PDP tailored to that company.

9.1 DEFINITION PHASE

The Definition Phase hasn't changed. The informal Wish List is written and used to convey marketing's input to the authors of the Functional Specification, which is the formal deliverable. The Definition Phase ends when the FS has been approved and released.

9.2 ARCHITECTURE/PLANNING PHASE

The Architecture/Planning Phase may begin after the Definition Phase is complete. It uses the FS as an input to drive the activities in this phase. The AS and Development Plan are the formal deliverables of this phase. When both have been released, this phase is complete.

9.3 HIGH-LEVEL DESIGN PHASE

The High-Level Design Phase may begin after the Architecture/Planning Phase is complete. The FS and AS are used as inputs to drive this phase, while the Development Plan drives this phase and the following phases. The HLDS is the formal deliverable of this phase; when it has been released, this phase is complete. The reason for making this a separate phase is to ensure that the AS, HLDS, and DDS are generated in sequence, allowing the opportunity to review each step before the next step is started.

9.4 Detailed Design Phase

The Detailed Design Phase may begin after the High-Level Design Phase is complete. The released FS, AS, and HLDS are all available to drive the Detailed Design Phase. Structured analysis, functional decomposition, or object-oriented analysis may be used as appropriate to derive the DDS from these documents [NASA1, p. 28]. The primary reason for separating the Detailed Design Phase from the Implementation Phase is to allow the conceptual design to be reviewed before implementation resources are spent. A classic study [YOURDON, p.105] found that it is an order of magnitude less costly to correct design errors in the Detailed Design Phase than after implementation. The DDS is the formal deliverable of this phase. When it has been released, this phase is complete.

9.5 Implementation Phase

In the Implementation Phase, the detailed design that has been documented in the DDS is implemented. This can be done quickly, because this design has been carefully planned, checked, and reviewed in the previous phases. As modules are implemented, they are unit tested, integrated into subsystems and tested as subsystems, integrated into systems and tested, then integrated to form the product and tested [NASA1, p. 9]. These tests don't duplicate the Acceptance Test, but are intended to identify, isolate, and remove bugs before the formal Acceptance Test begins.

The Implementation Phase may begin after the Detailed Design Phase is complete. This means that no part of the detailed, low-level implementation may begin before the driver of this phase, the released DDS, is available. The modification of modules from another product that are to be reused may not be started. If there is an obvious low-level implementation task that needs no formal input, it may not be started until the Detailed Design Phase is complete. If the implementers are idle, the implementation may not begin until the Detailed Design Phase is complete. If it makes absolutely no sense to wait, the implementation still may not begin until the Detailed Design Phase is complete. If the building is on fire, and the only way to put it out is to start the implementation before the DDS is released, the implementers leave the building and watch it burn as they work on the DDS.

The deliverable of the Implementation Phase is the design, generally embodied in a prototype unit, which is released to the acceptance testers after it has passed its review.

9.6 Acceptance Test Phase

The test engineers may start working to devise acceptance tests and write the ATP as soon as the FS is released. The Acceptance Test Phase may begin after the ATP has been released and the design has been released for test after the Implementation Phase. As the acceptance test progresses, an occasional test item may fail. In this case, the results of the test item are delivered to the implementers for resolution. If the implementers are able to correct the design by fixing bugs at the implementation level, they do so and deliver the amended design for re-test. If a change to the architecture and/or the high-level design and/or the detailed design is required to correct the defect, MR(s) are generated to define the change(s) to the AS and/or the HLDS and/or the DDS. After the MR is accepted, the change(s) are made, then the implementation is changed and re-tested. If a serious problem with wide-ranging consequences is found, the Acceptance Test Phase may be suspended while the project is returned to the appropriate earlier phase. Tasks of that earlier phase and all subsequent phases are repeated as needed (including reviews of all modified deliverables) to repair the defect, whereupon the Acceptance Test Phase is resumed. Test items are repeated until all pass. If some versions of the design pass some tests and other versions pass other tests, a regression test of the final design is done to ensure that a single version passes all tests. A Test Report is written to document this event.

The Test Report and whatever documentation is needed by the Production (Manufacturing, or Operations) Department to manufacture the product in quantity are the deliverables of this phase. By structuring the work in this way, the implementers can generate the production deliverables while the testers perform the acceptance test.

9.7 SUMMARY

The quantities of activities of the marketers, systems engineers, designers, and testers are summarized in **Figure 11**. During the Definition Phase, the marketers and systems engineers (the definers) work full time to generate the FS, while the designers perform feasibility studies, as discussed in Section 4.6.2 above. During the Design Phases, the designers get their turn to work, while the marketers use a very small fraction of their time to monitor the marketplace for unexpected changes that compel functional changes in the product. Then the others support the testers during the Acceptance Test Phase. This figure drives home the points that the design must not begin before the Definition Phase is complete, and that the marketers aren't working to find new features during the Design Phases.

Nice Chart!

DEVELOPMENT PHASES

Figure 11. Personnel Activity vs. Phase

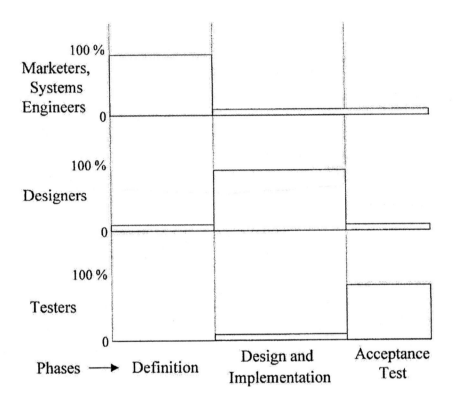

10

THE QUALITY PROGRAM AND CERTIFICATION

The company's Quality Procedures are the source of authority for the development program. While a full discussion of a quality assurance system is beyond the scope of this book, some points deserve attention here.

10.1 BASIC QUALITY PRINCIPLE

Every company tries to achieve a number of simultaneous goals, including:

- develop new products quickly
- develop new products at low cost
- manufacture new products at low cost
- attain high product quality.

Unfortunately these goals are frequently in conflict. If a particular action moves the company toward one of these goals but away from another, rules are needed to determine whether or not that action should be performed. Suppose for the moment that the rules are:

"All products must undergo final inspection, unless that would raise their production cost."

"All deliverables must pass a review, except when a review would make development slower."

With rules like these, nothing would ever be inspected or reviewed; the immediate gains in cost and time would always be allowed to outweigh the eventual intangible improvements in product quality.

If the quality program is to produce any benefit at all, clearly the priority rules must be very different from those above. If quality-related tasks are optional or conditional, excuses and rationalizations will always be found to circumvent their performance at the crucial times they are most needed. Therefore, the **Quality Procedures must be absolute and must have top priority**. Once a quality program has been put in place by management, its provisions cannot be waived or exceptions taken. This fundamental principle is not understood by some executives, who believe they have the authority to authorize departures from the quality procedures on the basis of expediency. When a low-level employee makes a mistake, it can be corrected at a review, or an MR can be used if the problem is detected after a deliverable is released. When an executive decision creates a violation of a quality procedure, there is no review to detect the problem, and the MR mechanism usually applies only to deliverables, not decisions. The Quality Assurance Department is the only line of defense in this situation; it must act promptly and decisively to ensure that the quality procedures are being followed completely in all circumstances.

10.2 VERIFICATION AND VALIDATION

The English words "verification" and "validation" have similar appearance, sound, and meaning. In quality programs, these terms have been given separate meanings, but these meanings are frequently confused because of the similarity of the two words.

"Verification" refers to procedures that are applied **during development** to determine whether the **development process** is being performed correctly. It investigates the deliverables of each development phase to find out if each deliverable is related in the proper manner to the references that are supposed to drive its generation. It monitors all of the activities that take place during development to see whether all of the procedures required by the PDP are being performed correctly in all cases.

"Validation" refers to procedures that are applied at the **end of development** to determine whether the **design meets its functional requirements**, as set forth in the FS. Validation is the goal of the Acceptance Test Phase described in Section 9.6 above.

Thus, verification tests the development process itself, while validation tests the result of that process [DUNN, pp. 136-138]. These are independent tests; either may be done without the other. In fact, many companies employ vast legions of test engineers to do validation, but no one at all to do verification! Believe it or not, it is the nominal responsibility of the design managers to do verification in their spare time. If just **one** of the test engineers were assigned to do real verification, many fewer testers would be needed to do validation, because there would be fewer bugs in the product. For the same reason, the overall time to market would be shorter, making the development less expensive.

In classic quality theory, verification and validation are seen as essential and **equally important**. The processes that were followed during development have been found to strongly influence the quality of products that are manufactured years later from the design that was developed. Clearly, this is a long-term benefit of using formal development methods.

Although verification typically receives much less attention than validation, it may be argued that verification is actually the more important of the two. After the development is complete and the product is being manufactured and sold in quantity, suppose the company wishes to improve its quality, perhaps because of customer complaints or high field failure rates. It is easy to upgrade the validation effort, by adding personnel or by devising production tests that identify weak units before shipment. But that product's development process cannot be upgraded, because it has been completed. Unless the product is redeveloped from scratch, the company cannot retroactively improve the development effort that yielded the product in question. For the rest of its life cycle, the product is blessed or burdened with the quality level inherent in its design. This is what is meant by the common expression, "Quality cannot be 'tested in'; it must be built in."

In summary, it is important to follow development procedures carefully and completely, because their effect follows the product (and all other products derived from it) for the rest of their lives. If tasks are done out of sequence or a review is skipped, the resulting weaknesses in the product can't be fixed later.

10.3 Reasons for Certification

Many customers are convinced that the quality of the components they buy is extremely important. If they intend to incorporate these components into larger systems and sell them, the quality of their products (systems) depends very directly on the quality of the purchased components. These customers are willing to pay a premium price for high-quality components, but want to be sure they get their money's worth.

Large, experienced customers usually understand that the high-quality components they seek can be produced only through formal development programs. They understand that functional testing of each production unit is insufficient; the details of the development process must be considered as well. To ensure that the purchased component is of high quality, the customer would have to validate the production of the component and verify the development process. An immediate difficulty is that the development may have taken place in the distant past. Another problem is that the customer may be using components from hundreds or thousands of vendors. If the customer's personnel had to thoroughly audit the processes of all of its suppliers (and all of their suppliers, and so on), they wouldn't have time to do anything else. If the vendor had to cooperate with a quality audit from every customer, the vendor's personnel wouldn't have time to do any development.

A solution has been found. An organization that is a recognized authority on quality can audit the processes of each vendor once, and certify those vendors that pass. The vendor needs to participate in only one quality audit, and prospective customers can consult the certification status of prospective vendors at very low cost. In practice, there are several organizations that are trusted quality clearinghouses, and periodic audits are performed to ensure that the integrity of the vendor's process is being maintained. Some certifying organizations will conduct an informal pre-audit to identify areas that aren't ready for certification, and help the firm upgrade those areas so that certification can be obtained on the first formal audit.

10.4 A Caution

One negative aspect of certification must be noted. Startups typically use informal methods to develop their initial basic product and its derivatives, for the reasons given in Section 13.3.11. Eventually the company makes the conversion to formal methods and gets its processes certified. It then uses formal methods as it develops additional product families; the Manufacturing Department uses certified manufacturing, test, and quality methods to produce all of the company's products. The company truthfully advertises that it uses certified formal development and production methods.

So far, so good. The problem arises because the company continues to sell the products that were developed informally. The company's product literature doesn't distinguish between products that were developed informally before certification was obtained and products that were developed formally afterward, giving the impression that **all** of its products provide the quality of products that had been developed formally. This situation lasts for years, until all of the informally developed products have become obsolete and their production is discontinued. In this way, unwary customers may receive informally developed products from a certified vendor, contrary to the customers' wishes. Customers can protect themselves by requesting a copy of the FS the vendor used to develop the product in question. (Another reason that the customer needs this document is given in Section 13.3.9.) If no FS can be delivered, or if the FS is inadequate to drive the formal development of the product, the true nature of the situation is revealed.

11

SOME ESTABLISHED FORMAL METHODS

This chapter lists a number of recognized formal methodologies that have been established for product development and briefly mentions their central principles. This material may help executives who are contemplating the selection of a formal method for their companies. It also emphasizes that the formal principles presented in the rest of this book aren't figments of the author's imagination, but are firmly established methods that are put to use daily by millions of practitioners. Some observations follow.

11.1 ISO 9001

The International Organization for Standardization (ISO) is the best-known worldwide standards body. It is supported by the national standards bodies of more than 120 countries. Its fundamental standard, ISO 9000, has been voluntarily adopted by **more than half a million** organizations. As of the end of 2001, more than 44,000 organizations had obtained certificates of conformity with its ISO 9001 standard, dealing with product development and manufacture. These numbers are rapidly increasing as companies find that their customers understand the advantages of formal development methods and prefer high-quality products. Numerous benefits resulting from the use of ISO 9001 have been quantified by surveys [ISO] and case studies; this material appears in Chapter 12 below.

The latest version of this standard, "ISO 9001 : 2000", abbreviated as "ISO 9001" herein, is intended for both product-oriented and service-oriented companies. The earlier (1987 and 1994) standards were expanded to apply to all types and sizes of organizations and industries. The previous 9002 and 9003 standards have been incorporated within ISO 9001. The companion ISO 14000 standards that deal with environmental issues are beyond the scope of this book.

ISO seeks to institutionalize a systemic orientation toward formal quality programs, so as to reduce the costs associated with poor quality. Organizations are encouraged to perform a Gap Analysis, to identify their gaps, or areas in which their present practice falls short of the level defined by ISO 9001. The obvious implication is that the organization should apply its resources to upgrade the deficient areas, so as to fully comply with the standard.

Section 7 of ISO 9001 requires certain procedures to be established and followed to guide the company's design and development activities. In addition to a general Quality Management Program, the following areas are identified as critical design and development operations that require formal procedures:

- Planning — a Design Plan is prepared and followed
- Inputs — the information that drives the design is obtained
- Outputs — the deliverables of the design process are used
- Reviews — of all outputs
- Verification — the design and development processes are monitored
- Validation — the design is tested functionally
- Change Control — functional changes are tracked and managed

11.2 CAPABILITY MATURITY MODEL

The Capability Maturity Model (CMM) has been developed by the Software Engineering Institute (SEI) at Carnegie Mellon University. Between 1987 and June 2001, more than 8,000 software development projects being run by more than 1,500 organizations have been examined and rated according to this model [SEI5]. CMM employs a "top-down" strategy, motivated by the belief that the quality of the development system governs the quality of the products it produces. It stresses a "common sense" approach to development, focused on both the product and the development process. In the form presented below, CMM applies only to software development. It appears that CMM's principles could be readily transferred to the development of electronic hardware as well as other technologies.

CMM [SEI1; SEI4] defines five **Levels** of process maturity, representing stages of software development capability. At each level, several **Key Process Areas** (KPAs) are identified, representing specific development procedures that have been found helpful by other developers in reaching the goal of that level. When an organization has demonstrated mastery of all of the KPAs associated with a given maturity level, it is promoted to that level. Levels must be attained in sequence, since each level is built upon the resources that were established in lower levels. The following sections discuss the five levels and their KPAs.

11.2.1 Level 1 — Initial

No matter how poor its processes may be, every organization starts at the Initial level; there are no KPAs required to qualify for Level 1. Companies that employ informal methods as described in Chapter 2, and "informal-plus" companies that write a few documents to "paper over" their informal process to give the appearance of formality, are typical of Level 1 organizations. Their process is characterized as *ad hoc*, leading to unpredictable performance.

11.2.2 Level 2 — Repeatable

The goal of Level 2 is to establish project management practices. The experience that SEI has collected over the years indicates that most project failures are due to management reasons rather than technical ones. Therefore, CMM builds a foundation of project management procedures and ensures that they are working properly before trying to put technical processes in place. The goals of Level 2 are to make the performance of the development process more predictable and repeatable. Six Key Process Areas are defined for Level 2:

- Requirements Management
  CMM assumes that the product's functional requirements have already been established and that some subset of these requirements have been allocated to software. The Requirements Management KPA addresses procedures for managing changes to these software requirements that take place after the design starts.

- Software Project Planning
  This KPA concerns the establishment of plans for doing the software engineering and managing the development project. Methods for generating schedules and budgets are covered.

- Software Project Control
  The Software Project Control KPA involves the processes for tracking the development of software, with reference to the plan generated in the Software Project Planning KPA. Management oversight into the actual status and progress of the development tasks allows detection of departures from the plan, and enables management action if needed.

- Software Acquisition Management
  This KPA applies to software development work that is purchased from a contractor as well as the purchase of off-the-shelf software and the use of software supplied by a customer. Processes for selecting and managing vendors of software labor are addressed. The other Level 2 KPAs are combined and applied to the vendor.

- Software Quality Assurance
  Through this KPA, management gains visibility into the processes actually used for software development and into the resulting product. A portion of the Quality Department is dedicated to monitoring the development process; this function may be separate from product validation.

- Software Configuration Management
  The purpose of this KPA is to maintain the integrity of the products produced by the development process. This task is in effect throughout the life cycle of the product.

11.2.3 Level 3 — Defined

The goals of Level 3 are to establish technical engineering practices and the support organizations that help them succeed, and integrate these tools with the project management processes built in Level 2. This level serves to institutionalize the technical and management practices across all projects throughout the company. There are seven KPAs associated with Level 3:

- Organization Process Focus
  This KPA serves to establish the responsibility within the company for process-related activities, with the goal of improving the capability of the processes.

- Organization Process Definition
  The Organization Process Definition KPA provides for the development and maintenance of software processes, to be accomplished by

the department or group designated in the Organization Process Focus KPA. The objective is to improve performance of the software development task, so as to obtain long-term benefits.

- Organization Training Program
  The goals of this KPA are to develop the skills and knowledge of employees so that they may effectively perform their tasks. The training activities that are related to product development are integrated with the company's other training operations.

- Integrated Software Management
  This KPA is an extension of the work started in the Level 2 Software Project Planning and Software Project Control KPAs. Its goals are to define coherent, company-wide processes for software engineering, project management, and risk management, and to provide for tailoring these standard processes to reflect the unique development environments and marketplaces of individual projects.

- Software Product Engineering
  This KPA contains the core of the technical processes for developing, integrating, and testing software, covering the project from the collection of requirements through maintenance. One output of this KPA would be a company-wide coding standard.

- Project Interface Coordination
  This KPA covers the proactive interaction between the software engineering group and other departments within the organization. Its processes serve to coordinate and manage these engineers as they communicate and interwork with other functional areas.

- Peer Reviews
  The Peer Reviews KPA defines the formal review process, as it applies to the software deliverables. The goals are early detection and removal of defects, and the training of others concerning the outputs being produced by the software group.

11.2.4 Level 4 — Managed

The goal of Level 4 is to establish quantitative management procedures to evaluate the quality of the products being produced and the **processes** that have been built in earlier levels. Its three KPAs are:

- Organization Asset Alignment
  This KPA calls for processes to foster planned reuse of software components, so as to obtain commonality across a product line.

- Organization Process Performance
  The Organization Process Performance KPA pursues the quantitative control of the performance of the development process. Baselines and models are used to explore tradeoffs.

- Statistical Process Management
  This KPA extends the Organization Process Definition, Integrated Software Management, Project Interface Coordination, and Peer Reviews KPAs. Quantitative measures are used to identify the causes of departures in these areas, leading to their correction and process stability. The software deliverables are measured to ensure that quality goals are met.

11.2.5 Level 5 — Optimizing

The goal of Level 5 is to constantly improve the development processes that have been put in place in the earlier levels. This work is built upon the quantitative measurements of the products and processes that were set up in Level 4. The three KPAs associated with Level 5 are:

- Defect Prevention
  The Defect Prevention KPA defines the process whereby the causes of defects are identified, with the goal of preventing similar defects from occurring in future products. An analysis of these defects may lead to an adjustment of the defined software process, as provided by the Integrated Software Management KPA.

- Organization Process Innovation
  This KPA defines the process whereby new technologies such as development tools and processes are found, evaluated, and introduced into the company.

- Organization Improvement Deployment
  This KPA provides for the distribution throughout the organization of the outputs of the Defect Prevention and Organization Process Innovation KPAs. Its goals are to continuously improve the software development process, leading to improved quality, increased productivity, and reduced cycle times.

11.3 NASA Standards

The National Aeronautics and Space Administration (NASA) has a number of working groups that seek to improve software quality, including the Software Engineering Laboratory, the Software Independent Verification and Validation Facility, and the Software Assurance Technology Center. The Software Engineering Laboratory has published detailed guides for software development [NASA1; NASA3] whose life cycle phases track very closely with those of Chapter 9. The entrance criteria of each phase match the exit criteria of the previous phase, so the phases are sequential, non-overlapping periods of time. The following development phases are recommended:

- Requirements Definition

- Requirements Analysis

- Preliminary Design (generate high-level software architecture)

- Detailed Design

- Implementation

- System Testing and Acceptance Testing

11.4 IEEE STANDARDS

The Institute of Electrical and Electronics Engineers, Inc. (IEEE) maintains a very large pool of standards in many areas. More than 40 of these IEEE standards apply to software engineering practices. In addition to "Guides" and "Recommended Practices", software-related IEEE "Standards" are available in the following areas:

- Configuration Management Plans
- Life Cycle Processes
- Maintenance
- Productivity Metrics
- Project Management Plans
- Reviews
- Risk Management
- Safety Plans
- Test Documentation
- User Documentation
- Verification and Validation
- Quality Metrics Methodology
- Quality Assurance Plans

11.5 GOVERNMENT STANDARDS

The United States Government has established an array of standards for contractors that develop products for federal and military customers. These standards are similar to those described above. Other countries have similar systems. If your firm is a government contractor, its contracts specify sets of standards and mandate their application. If not, the government won't designate standards that are appropriate for your firm or assist you in their use. In either case, no decision needs to be made; government contractors must use these standards and other organizations should not. For this reason, government standards aren't discussed further herein. Some aspects of the application of these standards is presented in [DUNN, pp. 206-211], along with an extensive case study [DUNN, pp. 254-279].

11.6 OBSERVATIONS

Some comments about these development methods follow.

11.6.1 Endorsement

The above standards or methodologies are not specifically endorsed or recommended by this book. All are worthy of consideration for adoption, in whole or in part. This book strongly endorses the use of any formal development method in preference to any informal method.

11.6.2 Similarities among methods

Clearly there is substantial overlap among concepts and terminology of these well-known methodologies. The researchers who composed each standard made use of common background material, but it is striking that the same principles have been found to provide significant benefits to organizations of many types, in diverse industries, operating in numerous countries and cultures, employing a wide range of technologies, and over decades during which substantial technological changes have occurred. The discovery that a given technique is necessary to ensure high quality in such an enormous spectrum of enterprises is a powerful validation of that technique. It suggests very strongly that this technique is also necessary in a particular company that may be considering its adoption.

This book attempts to abstract the common elements of various formal methodologies to derive the central principles of a generic formal development method. Attention is thereby focused on the importance of these principles, without the distraction of the detailed wording of any specific standard method.

11.6.3 General applicability

It appears that many of the above methods are oriented toward software-based products. But formal methods aren't just for software. In companies whose products integrate digital hardware and computer software that are jointly developed, it is essential that all development departments employ the same underlying methodology. Hardware and software (and electrical and mechanical) design should all start simultaneously at the beginning of the Design Phase, even when one department is expected to finish its tasks much earlier than the other departments. Architecture should be performed jointly, and architecture and high-level design documents should be reviewed jointly. After separate implementation according to individual technical procedures, hardware and software should be integrated and tested jointly by the designers before the start of acceptance testing.

Note that ISO 9001 was deliberately expanded to apply to the largest possible range of products, as well as services. A large body of practices are universally applicable, while detailed technical procedures specific to the technology involved are integrated within the overall structure.

In the software CMM model discussed in Section 11.2 above, only four of the 19 Key Process Areas are classified as Engineering activities; the remainder are regarded as Management or Organizational in nature. A company doing non-software development could replace those four software-related procedures with four equivalent technical procedures that describe its design processes, with only minor changes to the remaining KPAs. It seems a straightforward extension of the software CMM to expand the four Engineering KPAs to include whatever types of technology a company uses. The Software Engineering Institute is taking steps in this direction. CMM is being generalized to describe all forms of product development, while being extended to include the generation of product requirements under the heading of Systems Engineering.

11.6.4 Incremental migration toward formality

Under most formal methods, a company is asked to improve its performance in a number of designated areas simultaneously. If the company's resources for process improvement are small and the desired improvement is large, distributing these resources over all areas results in gradual improvement at best. It may appear that no significant benefits are being obtained, and that the formal approach isn't working and should be abandoned.

Improvement and benefits could be demonstrated if the resources for process improvement were concentrated in a small number of areas at a time, then moved to other areas after the initial areas had been brought under control. This is precisely the path that CMM has chosen. A few Key Process Areas are designated for initial attention. When the maturity of these procedures and activities has been established, the focus of improvement moves to other KPAs. But CMM doesn't just advocate working in a limited number of areas at a time; the **sequence** of activities is also important. **All** KPAs associated with a given Level must be completed before that Level is finished, and the Levels cannot be completed out of order. The management practices of Level 2 provide a foundation for the technical practices that follow in Level 3. If there is no mechanism in place to ensure that the technical procedures are being followed, further refinement of those procedures is probably wasted. Once management and technical procedures are in place, their results can be measured in Level 4 and then optimized in Level 5.

CMM provides a step-by-step sequential recipe for process improvement. Minor cumulative benefits should be realized as each KPA is accomplished, but the full benefits of formality cannot occur until the paradigm shift (see Section 14.2) takes place, and people voluntarily comply in the expectation of inherent rewards instead of going through the compliance ritual to avoid management's token displeasure. Since each CMM Level typically takes 24 months to achieve [SEI5, p. 28], improvement is a slow process. Dramatic immediate results shouldn't be expected.

11.6.5 Definition of a formal method

Section 1.2 promised that precise definitions of the terms "informal" and "formal" development methods would be stated in this chapter. A formal environment must have sufficient managerial and technical procedures in place to provide an infrastructure for rule-based development. An organization that has achieved CMM Level 3 (or equivalent) is capable of formal development. But a body of such procedures is inadequate by itself. Procedures must require the equivalent of a written FS, and must require the approval of this document before design begins. These procedures must require the design to comply with the FS. These procedures must be followed in all cases. If all of these elements are present, formal development is taking place. If not, for the purposes of this book the environment is defined as informal.

Many executives interpret the various standards and models presented in this chapter as definitions of **ingredients**, whose use results in formal development. This error leads directly to an "informal-plus" methodology. To achieve successful development, the **recipe** portions of the standards and models are also needed to guide the combination of these ingredients to form a fully integrated formal process. Writing, reviewing, approving, and archiving formal documents and then ignoring them is equivalent to providing separate measured quantities of flour, sugar, milk, and eggs without combining or heating them. A procedural recipe is needed to specify how the ingredients are to be combined and processed to become a product or a cake. It should not be assumed that employees who are provided with ingredients will know how to process them, or even recognize that part of the recipe is missing. The company's Product Development Procedure (see Section 7.1) must specify both the ingredients and the procedural recipe, not just provide a list of ingredients. This point deserves emphasis because ingredients-only attempts at "formal" methods are all too common.

12

BENEFITS OF THE FORMAL PROCESS

The reason for adopting a formal development process is to obtain the numerous tangible and intangible benefits that result. This chapter lists many of these benefits and relates them to specific formal procedures and documents where possible. These benefits are defined by contrasting the results of a generic formal method with those of a typical informal method. Although many benefits can be described only in qualitative terms, in some important cases quantitative measures of actual experience are presented.

Many benefits are derived from the single fact that a robust FS is approved before the design begins. Once a formal FS has been constructed to solve the communication problems described in Section 4.3, many additional users emerge throughout the company, and other benefits come raining down upon these users of the FS. If any single benefit were the only reason for generating the FS, the cost of preparing a high-quality specification might be hard to justify. However, when the cost is shared among numerous users and purposes, it's easy to justify the investment. The availability of a competent FS enables many uses that wouldn't be possible if the FS had to be generated "from scratch" for each purpose alone. As new uses are conceived, the incremental cost of the FS is zero for each new use. It's easy to find uses for a resource with zero cost. Other formal documents also contribute benefits in other ways. Before listing these specific benefits, a few general observations are in order.

The development programs of large companies tend to be more formal than the programs of smaller firms. Certainly it's easier to informally control the efforts of a small number of people than a sprawling project involving thousands of developers, so large companies are driven toward formality out of necessity. Perhaps this isn't the only reason for the difference in formality. Perhaps evolution by natural selection is at work. Could it be that some small firms adopt formal programs that help them grow into large companies, while other small firms remain informal, and stay small as a result? Are you willing to bet that this factor isn't at work in your industry today?

One of the most powerful pieces of evidence of the superiority of formal methods comes from a survey that the Software Engineering Institute conducted among organizations that were increasing the maturity of their software development methods, as measured by the Capability Maturity Model (see Section 11.2). A report [SEI2] of the results of the survey is called the "SEI Report" in this chapter. A total of 13 diverse organizations participated covering a wide range of commercial and governmental applications, including information systems, telecommunications, operating systems, and embedded real-time systems. As these organizations improved their CMM ratings, substantial benefits were measured. Other studies [SEI3] confirm these results.

Another study that supports the benefits of formality was conducted by NASA. A report of their results is called the "NASA study" [NASA2] in this chapter. Various studies [ISO] have been made of ISO-registered organizations to quantify the extent of the benefits that resulted from formality.

In a number of the following sections, it is suggested that the FS could be shown in various ways to people who aren't covered by the routine nondisclosure clause that has presumably been signed by all of the company's employees. A carefully prepared FS for a complex product represents a substantial capital asset, which must be protected from access by potential competitors. In all cases, it is assumed that appropriate legal nondisclosure agreements are obtained before the FS is disclosed. This assumption isn't restated in the individual sections to which it applies.

In a number of the cases discussed below, the opportunity to use the FS may arise suddenly, with a time horizon too short to allow a competent FS to be written "from scratch". Many of these opportunities originate from events external to the company and outside its control. Since the FS is written in the first phase of the development process, it will be available whenever it is needed to capitalize on each opportunity.

In order to organize these benefits, most of the benefits are attributed to a single department within the company. This association is somewhat arbitrary, in that many benefits affect several departments. In a larger perspective, whatever helps one department helps the entire company. Before explaining the benefits that the FS brings to individual departments, some benefits that accrue to the company as a whole are examined.

12.1 Company-wide Benefits

If the FS were destroyed at the moment of its approval, the process of generating it would have benefited its authors and reviewers by focusing and organizing their thinking about the product's functions. Much larger additional benefits can be obtained all through the life of the product by using the FS as a powerful tool.

12.1.1 The FS communicates a uniform definition to all areas

This is just a statement that the FS actually accomplishes the objective of solving the problems presented in Section 4.3. If the FS is stored as a computer file on a server that all authorized employees can use, access to the FS is almost instantaneous. The file is updated whenever a new FS version is approved, so that current information is always available.

Having the FS in machine-readable form allows it to be searched by a word processor utility to rapidly locate all references to a given topic, an advantage over hard copy.

12.1.2 Documents store knowledge over time

The FS stores an unchanging detailed functional description of the product, except for deliberate modifications made under change control. It records the results of the decisions and compromises made during the Definition Phase and the approved changes that accumulated thereafter. Its memory is clearly superior to that of a collection of people. An individual may author specifications for a sequence of very similar products. While she is specifying Model 4, she can't be expected to recall with certainty all of the details of Model 2 that was finished eight months earlier.

If the people who invented and championed a product leave the company or move to other assignments, the FS and design specifications store their work, providing continuity for the new employees who maintain, modify, manufacture, and sell the product in the future. In short, the formal documentation removes the company's dependence on specific people's memories of the product's functions, architecture, and implementation.

12.1.3 Documents are easily transported

Companies are increasingly making use of remote development and manufacturing centers. Such a facility may be established to obtain lower labor cost, or a company may acquire a subsidiary at a remote location. Any of the formal development documents can be sent quickly and cheaply anywhere in the world by secure email, to coordinate the work taking place at the remote centers.

In an informal system, the information is stored in the memories of numerous people. The difficulties of accessing this information, as presented in Section 4.3 above, are compounded if some of the information sources work at remote locations in different time zones. If large quantities of information are needed at a remote center, usually one or more people must travel there. Certainly the transmission of a document is preferable.

12.1.4 The FS makes specification changes visible

"Visibility" sounds like an abstract term, useful for passing exams in business colleges but with little practical application. However, the visibility that is provided by an FS allows management to detect changes in the product's specification that would be overlooked in the absence of a formal document. This is such an important point that the following (fictional) account has been prepared to illustrate the extreme damage that can result when visibility is lacking.

The Parable of the Wheels

Consider a company that designs, makes, and sells microwave ovens using informal methods. A salesman phones the marketing manager (let's call him Ralph) and says excitedly,

"I just talked to an appliance wholesaler who wants to use his oven in his kitchen to prepare supper, then move it to his den so he can heat his TV snacks without leaving the TV area. I think it would be great to put wheels on our product!"

Ralph likes this idea very much. The concept of moving an oven has never occurred to him; it's a totally new, out-of-the-blue breakthrough. It sounds like a differentiating feature, the sort of thing Ralph is being paid to discover. However, he knows that if he approaches management with this concept, the salesman will take credit for it, the concept will be put on the Future Feature List, and it will be 18 months before the first wheeled oven can be delivered. If Ralph demands that wheels must be added to the oven presently nearing design completion, the engineers will throw a fit, and Ralph will be blamed for delaying the new product's delivery. To retain credit for his wheels idea and avoid blame, Ralph decides to do some "Internal Marketing".

BENEFITS OF THE FORMAL PROCESS 145

To plant the seed, Ralph approaches the designer who did the mechanical drawings for the oven under development. Ralph says,

"I loved the prototypes of your design, but I didn't see any unit with the optional wheels. I just stopped by to re-confirm that the mobile unit will be ready on schedule."

"Wheels??"

"Sure. You remember the Product Planning meeting, when we talked about using the oven in the kitchen or the den…"

"We understood a fixed installation, in one room or the other."

"Well, we have a lot of customers who want to move their ovens between rooms, that's why we included the wheeled option. By the way, market research indicates that 50 feet is the right length for the retractable extension cord."

"What??"

"If you don't work on anything else, how soon will the mobile unit be ready?"

"Oh, I guess we could throw on some wheels in a month or so."

While the designer looks for his supervisor, Ralph travels in a straight line at the speed of light to the company President.

"I've got bad news, Boss. The new oven will be a month late. You remember the planning meeting, when I asked for a unit that could move between the kitchen and the den? The model we're planning to sell as the Rovin' Oven?"

"Well, I remember hearing about use in the den…"

"That's right. I still can't believe it, but somehow the Engineers forgot the wheels!"

Now Ralph plays his trump card.

"In that Sales Forecast I gave you last month, naturally I assumed the mobile version would be available for sale. Without wheels, we can sell only two-thirds as many."

In order to hold Ralph accountable for his Sales Forecast, the President must support every assumption that Ralph alleges led to that estimate. The President summons his engineering managers.

"You blankety-blanks forgot the blankety-blank wheels! You always do something like this to me. Are you telling me that none of those drawings you've been reviewing for the past six months showed any wheels, and none of you caught that!? It's a good thing

Ralph isn't asleep at the switch! You have three weeks to finish the mobile model."

No matter how loudly Engineering protests that a change of scope occurred, it looks like they're just trying to cover their oversight. Of course, forcing a substantial change late in the development process produces an ugly kludge. Sales of the new oven are negligible. No one wants to operate his oven at ankle height or trail a cord from room to room. Wheels don't work on stairways. Even the original "customer" isn't interested; all he wanted was a handle on top of the oven. As a result, the VP of Marketing is fired, and Ralph is promoted to her position. It turns out that Ralph wasn't the champion of the disastrous wheels after all; they were the idea of some lunatic salesman. Ralph just mentioned the idea to the engineers, who went crazy and wrecked the product. They should have known better than to make a large change so late in the development cycle. (Most of the engineering staff now works for more honest employers.)

Let's analyze what happened here. The specification of the new product changed, in a sudden, unexpected, and significant way, and management didn't acknowledge that any change occurred. This is called a "stealth change". There was no opportunity for estimates of recurring and non-recurring costs and schedule slips, including impacts on other ongoing development projects. The adverse market factors were never considered. Management didn't get the opportunity to evaluate the costs and benefits of adding a mobile option and reach a rational decision. Instead, Ralph unilaterally subverted the product planning process to his own ends, based on a snap decision. Left unchecked, several stealth changes can impact each new product, destroying the development process.

What would have happened if this oven company had based its developments on detailed, written, approved, controlled specifications? Since the concept of moving the oven is new, there would be no mention of relocation in the approved FS. When wheels were proposed, it would be obvious to all that a significant change was involved. This would have flagged management to follow the formal procedure to modify an FS, ensuring that due diligence was performed before approving the modified FS. With a formal process in place, the subterfuge wouldn't have been attempted.

12.1.5 The FS motivates work at the appropriate phase

The FS is the vehicle that stores the result of the Definition Phase for use by all later phases. Its presence focuses attention on the customer-related activities of the Definition Phase. If the customers' needs aren't fully explored during the Definition Phase, this fact is revealed by missing or fuzzy FS sections or open issues. While the Wish List serves as the Marketing Department's mid-term exam, the FS is its final exam. The formal FS review meeting allows executives from all departments to form opinions about the adequacy of Marketing's effort and its performance, based on the quality of the FS. This public exposure motivates Marketing and other team members to exert their best efforts during the appropriate phase.

12.1.6 The FS minimizes changes during design

An FS is a commitment made by its authors, reviewers, and approvers that the company is prepared to design, manufacture, sell, and service the product exactly as it is defined in the FS. The Product Development Procedure and document control system deliberately make it difficult to change the FS after it has been approved. Management has explained that a major goal of the formal system is to reduce the number of functional changes that are made during the Design Phase. Management has charged the MR Reviewers with the duty of carefully screening the MR's, accepting only the essential changes. Thus, a stigma is attached to any attempt to make a late functional change. After several proposed changes that would have taken place in the informal environment are rejected by the formal system, the employees realize that management is serious about its commitment to the principles of the formal program it has adopted. The lack of opportunities to make late changes motivates employees to concentrate their functional creativity during the Definition Phase of the next product's development. In a more positive vein, management should reward its employees when a successful product development is completed without significant functional changes.

12.1.7 Formal methods contribute to higher product quality

The evidence that quality programs improve product quality is overwhelming, but outside the scope of this treatise. The SEI Report [SEI2] presents quality improvement in terms of the rate of reduction in post-release defect reports. This refers to defects found by customers, so it's a very important statistic. The median organization showed a compounded yearly reduction rate of 39 per cent, an extremely impressive figure. The NASA study [NASA2, pp. 43-47] confirms this finding, reporting a 75 per cent reduction in the rate of defects over an eight-year period. Eighty-five per cent of ISO-registered firms reported [ISO] that customers perceived higher quality in their products, resulting in greater customer demand. In another study [SEI3, p. 13], almost two-thirds of the organizations at CMM Level 3 rated their product quality as excellent, while only 8 per cent of organizations at CMM Level 1 said their products were excellent.

12.2 HUMAN RESOURCES-RELATED BENEFITS

The formal documents provide benefits in several ways that are employee-related, without being directly associated with an operational department.

12.2.1 The FS as a training textbook

Many firms conduct training classes for their employees and customers on the functions, operation, and maintenance of their products. The FS, or material derived directly from it, can serve as a textbook for such a class. An advantage of using the FS itself is that the document is kept current, as described in Section 4.8.5, without any effort on the part of the training staff.

12.2.2 Rapid learning by new or transferred employees

One significant benefit of formal documentation is often overlooked. When new employees are hired or existing employees are transferred onto a development project, their first task is to get up to speed on the objectives of the project. A well-written FS is an excellent teaching tool, since it contains exactly the information the new employee needs, and no other obscuring material. By studying the FS, an accurate, current, common understanding can be quickly achieved, without burdening the employees who have been working on the project with tutoring duties. Once the FS has been understood, other formal documents, such as the Architecture Specification, High-Level Design Specification, and the Development Plan, can be studied (even in draft form) in preparation for the new employee's specific duties. The availability of these documents increases the productivity of the new employees by allowing them to promptly start performing these duties upon a solid factual basis. The productivity of existing employees is preserved by not diverting them greatly from their tasks to answer fundamental questions.

12.2.3 The FS as a recruiting tool

Every company tells candidates for employment that it maintains a set of adequate Quality Procedures that are carefully followed to guide its development process. Companies that actually behave in this manner have an FS for each product that has completed its Definition Phase. They can gain credibility by exhibiting such an FS for brief inspection by candidates, without disclosing trade secrets. The candidate can verify the existence of the document, and assess whether it is complete, detailed, well-organized, and expressed in formal specification language. An experienced user of functional specifications can make this assessment quickly, and will be favorably impressed by a robust FS. A person who has worked productively in a formal environment elsewhere regards the FS as a fundamental tool, absolutely necessary for the performance of his or her job. To such a person, it would make as much sense for management to forbid employees to use desks, forcing them to do all their work standing up, as it would to deprive them of the essential FS.

On the other hand, if a candidate expresses disinterest in the FS and seems unfamiliar with formal development environments, the reviewer records a significant negative factor for this candidate. Evidence of hostility to disciplined methodologies should disqualify the candidate.

(Help-wanted ads frequently seek an employee who can "follow a product from conception to delivery". This is a strong clue that the company uses informal methods. The employee must carry all of the product-related information in her memory throughout the development cycle and perform all development tasks personally. If she were allowed to use robust documentation, she could hand off the product to definers, architects, designers, or test specialists at the appropriate times, but the company doesn't intend to operate in this way.)

12.2.4 The FS helps retain employees

In Sections 4.9.3 and 12.1.4, we see how publishing a detailed FS prevents unscrupulous employees from camouflaging the capricious changes they make to a product being developed, so that they can pretend that no change occurred. The same theme appears elsewhere in various forms in this book; the FS serves to defend the productive employees who are following the company's procedures from manipulation and unfair blame transfer from unprincipled colleagues. In an uncontrolled environment, even at times when no changes are being made, the constant threat of future stealth changes overhangs the workplace. By providing protection against such abuses, the presence of a formal process removes a major negative factor from the environment, helping to retain employees who value fairness and honesty.

A developer who has worked in both informal and formal environments may well say, "To me, the ultimate horror is to try to implement something without having a competent FS in place. My employer's formal Quality Procedures are my Bill of Rights; they are management's pledge that I will never be forced to participate in an amateurish project ever again. I place a high value on this guarantee."

Among ISO-registered firms, 50 per cent reported [ISO] improvements in employee motivation. Organizations reported [SEI3, p. 13] that their employee morale improved as the maturity of their processes increased.

12.3 Management and Finance-related Benefits

An FS even provides benefits directly to top management and the Finance Department. One benefit is that the very existence of the FS communicates a sense of professionalism to casual readers who may not understand the technical details it contains. If the FS is competently organized and written, readers familiar with formal development practices will be favorably impressed. On the other hand, if the FS is poorly constructed and sloppily written, those readers will be negatively impressed with those responsible for the document. This benefit is in addition to the specific advantages listed in the remainder of this section.

12.3.1 Communication to Top Management and Directors

The FS can be used by the President (or CEO, COO, etc.) to learn the details of the products his or her company makes, in order to make informed decisions and speak intelligently to stockholders. Direct access to the FS removes the layers of filtering in the ordinary corporate communication channels that may obscure important information in some cases. Executives remember that certain features have been discussed in connection with the product; reading the FS reveals which features have actually been selected for inclusion in the current release.

Corporate Officers and Directors may inspect the quality of the FS to assure themselves that an important part of the corporate Procedures they authorized is actually being carried out.

12.3.2 Communication to potential investors

Suppose outside investors are studying the company to decide whether to do a friendly takeover. To learn the staff's level of competence and dedication to following its formal procedures, and to gain a detailed knowledge of the functions of the company's product, the investors (and their technical support personnel) could interview a number of the staff members. Or they could get a copy of an FS from the company's management and study it. In addition to convenience and accuracy, the FS has another key advantage: it cannot leak the information to someone else that a potential takeover is "in the works". Premature disclosure of an impending acquisition could allow speculators to buy the company's stock, raising its price and making the acquisition more costly to the investors. Therefore, the privacy afforded by the FS is a significant benefit in takeover situations.

12.3.3 Value to potential investors

To potential investors who understand the benefits described in this chapter, a company with a formal development method in place appears much more valuable than a company that develops informally. This difference may determine which company is selected for acquisition. It would be relatively easy to integrate the formal company into the formal environment typical of large acquiring firms. (Many large companies want all of their divisions and subsidiaries to do certified formal development, so that the company can claim overall certification.) The new owners of the informal company would have to pay the costs (in money and time) of attempting a transition to formality; if there are many laggards embedded in the acquired company, this transition may be greatly delayed or may fail.

12.3.4 Sale or transfer of a product being developed

An organization may start to develop a product, and then wish to sell the partially developed product to another firm or transfer the development to another division of the company. This might be done because of a redirection of corporate goals, as part of a restructuring, to transfer a proficient design team in its entirety to pursue an emerging "hot" opportunity, or to raise cash.

At the end of each formal development phase, the information that has been generated and the decisions that have been reached during that phase have been recorded in the deliverable(s) of that phase. These documents have been reviewed, corrected, and placed under document control. Although some "look-ahead" activities may have been performed, the formal work for the next phase hasn't started. Thus, almost all of the value generated by the developers so far is preserved in the deliverables of the phases that have been completed. The end of a phase is an ideal opportunity to sell or efficiently transfer the development project, which is done by transferring a package consisting of these deliverables.

Transfer is much harder if the development has been conducted informally. Some of the information that has been generated so far is in the form of scattered "hard copy": notes, meeting minutes, partial informal documents, sketches on scratch paper, emails, memos, and letters. The material that reflects the current thinking of the developers isn't differentiated from obsolete approaches. Even if all such material could be retrieved and transferred, it is collectively insufficient to convey all of the information needed to carry on the development. Much of the crucial information resides in the minds of the key personnel who have been working on the project. To fully transfer the development implies the transfer of these people. If the project is sent to a geographically remote division of the company, all of these key people would have to relocate. If the project is sold to another company, all of the key people would have to become employees of the acquiring firm and relocate to its facility, which might involve immigration to a foreign country.

12.3.5 High return on investment

The SEI Report [SEI2] defines "business value" as the ratio of savings to the cost needed to produce that savings. The median business value for the organizations that reported this statistic is 5.0, and the minimum is 4.0. That is, money spent on increasing the maturity of the development process leads to a savings of four or five times the amount spent. There are very few investments that are this lucrative. Since the calculation of savings ignores important intangible factors, the actual return is somewhat greater. The SEI Report [SEI2] revealed a median expense of $1,375 per year per software engineer, roughly one per cent of the loaded cost of employing that engineer. NASA reports [NASA2, p. 49] that a routine process improvement program costs less that one per cent of their total software budget.

12.4 Manufacturing Benefits ☆

The FS provides significant benefits to the department that manufactures the product it describes. For the purposes of this discussion, it is assumed that this department includes Test and Quality responsibilities.

12.4.1 The FS enables a quality program

The FS formally documents the functions of the product as seen by its user. It provides an objective, quantitative reference against which the actual product can be compared, thereby enabling a formal Quality Program. In the absence of such a standard, quality cannot be ascertained; a function can be observed, but not tested by comparison with a reference [DUNN, p. 161]. One executive may observe the operation of the product under given conditions and say, "It looks OK to me", while her colleague observes the same functional behavior and says, "That's not what I expected". The ensuing discussion may determine the desired functionality, but it's too late; one variation has already been implemented, so rework and delay will occur if any other variation is selected now.

Ninety-five per cent of ISO-registered firms reported [ISO] a greater employee awareness of fundamental quality principles, leading to increased operational efficiency.

12.4.2 The formal documents provide feature traceability

In serious quality programs, all of the design documents are required to reference the FS. The description of each device, circuit, and software module must state which specific section of the FS it implements. A Compliance Matrix can be constructed from this information, showing which section(s) of the design documents address each FS section

[YOURDON, p. 168; NASA1, p. 26]. This tool allows design reviewers to determine which FS Sections haven't been addressed.

If a mechanism, circuit, or software module isn't associated with any identifiable FS Section, it isn't needed, so it must be removed from the product. This guards against "feature creep", as discussed in Section 4.9.5.

12.4.3 Potential for quality certification

At some time, the company may want to have its Quality Program certified by a recognized authority such as ISO or ANSI. Customers may demand certification, or the company may want guidance in improving the quality of its products. In order to get certification, the company's Procedures will have to provide for a written functional definition of its products. Generating and using the FS thus serves as preparation for eventual certification. Experimenting with a formal program before certification is needed provides valuable practice, allowing the company to fine-tune its Procedures based on its individual experience. The certifying authority will require the company to operate for some time under its formal program, to demonstrate that it actually follows its Procedures carefully. If eventual certification is on the horizon, the sooner the company starts its formal program the better. Certification is addressed in Chapter 10.

ISO registration leads to increased international recognition and product acceptance; 69 per cent of registrants reported [ISO] that registration gave their firms a competitive advantage in the marketplace.

12.4.4 The FS enables a robust, predictable Acceptance Test Plan

There is a simple rule that leads to the proper Acceptance Test Plan:

TEST TO THE FS

The FS contains a complete statement of the product's functionality as seen by its user, so there is no need to look elsewhere for requirements to be tested. Features that aren't in the FS must not be included in the product.

It's important for designers to be able to predict what the product needs to do to pass its acceptance test, so as to provide the necessary features, and no others. The FS is the only mechanism that provides this coordination. The testers generate their Acceptance Test Plan according to a strict rule: every statement of material fact concerning the product's functions in the FS leads to a test item. No other tests are to be performed. The designers know that this rule will be used to generate the test, so they apply the identical rule to the FS to determine which features they should design.

12.4.5 Other Manufacturing benefits

Among ISO-registered firms, 20 per cent reported [ISO] an increase in on-time delivery, 53 per cent had less scrap and waste, 40 per cent reported a reduction in manufacturing cost, while the average defect rate was reduced from 3 per cent to half of one per cent.

12.5 MARKETING AND SALES-RELATED BENEFITS

The FS is produced during the Definition Phase. The Marketing Department concentrates its efforts during this phase, and is rewarded with a document that provides a multiplicity of marketing-related benefits throughout the product's lifetime. The availability of an FS relieves Marketing from having to repeatedly describe the product's functions to different groups needing functional information. Thus, marketing personnel can spend their time productively, performing marketing activities for other new products that are in their Definition Phases.

12.5.1 Formal methods can reduce Time To Market

The SEI Report [SEI2] shows a median compounded annual rate of reduction in time to market of 19 per cent. As a firm's development process matures, products are developed **faster**, contrary to prevalent belief. This is primarily due to reduction in false starts, rework, and functional changes. A shorter development time is of great importance to marketing in its quest to beat the competition to the market.

The SEI Report shows that some organizations can achieve remarkable acceleration of their development programs; it doesn't guarantee that all organizations can do so. But time to market has become so important that the possibility of a significant reduction should motivate all companies to give serious consideration to formal methods. If the median rate of 19 per cent per year can be compounded for three years, time to market will be **cut in half!** If a company can achieve the minimum rate shown in the SEI Report of 15 per cent, it will need four years to cut its time to market in half. Clearly a shorter development interval implies a lower non-recurring development cost, and a higher rate of return for the product overall.

Case studies [ISO] have found that ISO registration resulted in a 40 per cent **reduction** in development cycle time.

12.5.2 The FS separates WHAT from HOW

Engineers are notoriously poor writers of functional specifications. They are so entranced with technical details that they cannot resist explaining HOW the product will work when they try to write a functional description. Implementation issues may need to be addressed during the Definition Phase, but they are out of place in the FS, a functional document.

The rigid separation of WHAT the product does from HOW it operates has proven to be a very powerful tool for organizing the thinking about both issues. (Section 6.1.1 treats this matter.) By specifically prohibiting implementation issues from the FS, Marketing compels everyone to focus on its primary concern: WHAT does the customer want the product to do? Getting the right feature set is crucial. Nothing is worse than a product that's not welcomed by its prospective users.

12.5.3 The FS is a reference for salespersons

Salespeople have a simple rule to follow:

SELL THE FS

In the FS, the salespeople have access to a complete functional definition of the product, expressed from the user's point of view, and unencumbered by implementation details. The current FS is a complete representation of the company's present intentions with regard to a product's functions, which is updated when those intentions change. This is an ideal reference for salespeople to study, to learn the product's capabilities, to assist them in preparing sales presentations, and to answer customers' functional questions. Salespeople may consult whatever references they please to learn about a product's non-functional characteristics (price, availability, warranty, return policy, etc.), but they must rely entirely upon the FS for functional issues, to avoid creating a sneak path.

There may be circumstances in which marketing management decides to promise a prospective customer that a feature that isn't under development will be delivered soon, or exaggerate the capabilities of a product, or conceal a weakness. The presence of the FS ensures that their tactical decisions aren't made out of ignorance of the actual features of the product.

12.5.4 The FS allows specific change proposals

The FS serves as a baseline. Proposed changes to this baseline can be made specific by referring to numbered FS Sections. In the absence of a detailed baseline, it's hard to formulate a specific functional change, so requests are expressed as general complaints, or as problems to be fixed, without a specific solution being proposed.

12.5.5 The FS can be shown to selected customers

A serious prospective customer may need more functional information than is in a sales brochure. He may want to determine whether the product will interoperate with his present equipment, or to see if his technical personnel have the skills to install, configure, operate, and maintain the product. In addition to supplying this information, a competent FS impresses the customer and demonstrates that a very important part of the formal quality program promised by the vendor is actually being performed (see Section 10.4). In addition, the salesperson's time is saved if the FS can answer questions that the salesperson would have had to look up or ask Headquarters to answer in the absence of an FS. The FS is likely to provide answers that are more consistent and correct than the salesperson can provide. Section 13.3.9 discusses a different reason that customers need access to the FS.

In a different scenario, a customer may need to generate a functional specification to be used to solicit bids from multiple vendors. To save time and effort, the customer may informally collect FSs from several potential vendors, and select the best FS to serve as its baseline. This baseline is changed to include the unique features needed by the customer, and distributed with the bid package. Clearly, the vendor whose

FS was selected as the baseline has a significant advantage over the other vendors; the required functions depart only slightly from the standard product defined by the FS, while competing vendors must make substantial changes to their standard products in order to comply. Having a competent, complete FS that is suitable for inspection by customers is a prerequisite for obtaining a competitive advantage in this way.

12.5.6 The FS can be submitted with a bid

When a customer solicits proposals that are to include technical details, the FS can be include with the proposal to satisfy this requirement. If some part of the existing product is inappropriate for this particular application, a responsive proposal can be written by starting with the FS as a baseline. This baseline is modified to describe a customized product, similar to the standard product described by the FS, that meets the customer's needs. Since the FS has already been prepared, a proposal based on it can be generated in a very short time. If no FS were available, the vendor would have to do research to characterize the standard product, and then take significant time to draft, review, and polish the technical proposal.

12.5.7 The FS can be compared with a customer's specification

If the customer distributes a functional specification and asks vendors to submit proposals to supply a conforming product, the customer's specification can be easily compared with a number of FSs that define the vendor's product line, to quickly determine which standard product is closest to the desired article. The differences between the customer's specification and the closest FS are the modifications that must be done to the corresponding standard product to satisfy the customer's requirements. This list is very helpful in estimating the scope of the customization effort, and hence the amount of the vendor's bid.

12.5.8 The FS can be shown to distributors or sales representatives

In much the same way that the FS was used as a personnel recruiting tool in Section 12.2.3, it can be employed to educate, impress, and help recruit distributors and sales representatives for the product the FS defines. These representatives may be more willing to work with a company that is able to supply competent documentation than one that cannot.

12.5.9 The FS is the basis of the User's Manual

The FS states the functions and features of the product as seen by its user, without being cluttered with implementation information and extraneous material. It contains all of the details that may be of importance to users, and is logically arranged. Thus, the FS is the ideal reference from which to write a User's Manual, intended to instruct a user who is unfamiliar with a product in its operation. Further insight can be obtained from the FS authors and reviewers who have been playing the role of the user, as described in Section 8.5.1.

The FS is available early in the product's development process, allowing the writer of the User's Manual to start organizing information and preparing drafts of the Manual at the start of the Design Phase. Since a great effort is being made to minimize changes during development, only a small quantity of rework should be necessary. By contrast, the writer in an informal environment must wait until the Design Phase is nearly complete before starting the Manual effort, because the User Interface has been mutating unpredictably throughout the design process. The quality of the Manual suffers if it must be produced hurriedly.

12.5.10 The FS is the basis of sales brochures and specification sheets

The FS is also an excellent reference for the writers of Sales Brochures and Specification Sheets. The FS can be summarized to produce the Specification Sheet, which can in turn be condensed to yield a Sales Brochure.

12.5.11 The FS is the basis of advertising copy

The FS serves as a reference for writers of ad copy, particularly technically oriented ads that stress features, compliance with standards, and interoperability.

12.5.12 The FS communicates to a marketing consultant or business planner

When the Marketing Department retains the services of a Marketing Consultant or Business Planner, the FS can be used to inform that person of the functions of the existing products and those that have completed their Definition Phase.

12.6 CUSTOMER SERVICE-RELATED BENEFITS

Having a detailed FS allows Customer Service to answer two fundamental questions before they begin "repairing" a product in response to a customer's complaint. Is the product working as specified, or is it broken? Is the product being operated in its specified environment? Unnecessary service calls can be avoided by consulting the FS to help answer these questions before maintenance is authorized.

Seventy-three per cent of ISO-registered firms reported [ISO] improved customer service; in a case study [ISO], registration resulted in a 30 per cent reduction in customer claims.

12.7 ENGINEERING-RELATED BENEFITS

The benefits to the Engineering Department are presented last, to emphasize the point that the FS isn't just for the designers. As shown above, nearly everyone in the company derives some benefits from it; in many cases, these benefits are substantial.

12.7.1 Designers know what to design

Designers are the most immediate and intense users of the newly approved FS. During the Design Phase, they obey one simple rule:

IMPLEMENT THE FS

According to the Fundamental Rule, there is no need for them to look elsewhere to find out what the product is supposed to do. Professional developers are able to quote from memory the sentences of the FS that specify the part of the product they are currently designing. They know whether the FS says "and" or "or", and they have developed a rational interpretation of the requirements that guide their everyday work. They didn't set out to memorize anything, but they referred to their copies of the FS so constantly that they learned them automatically.

When a change is made to the FS, the same rule applies. The designers continue to comply with the old FS, ignoring the new FS until it is officially promulgated. Then they reorient their efforts toward implementation of the newly-approved FS. (Practical measures to avoid wasted effort are presented in Section 4.9.4.)

The situation is quite different in an informal system. Two managers of such a company were actually overheard to say, "What are we supposed to be building?" and "Nobody has told me that's a requirement". The lack of direction of the first manager is painfully apparent; what sort of instructions do you suppose he gives his subordinates? The second manager is working from verbal inputs, a hallmark of an informal system.

12.7.2 Designers know when to stop

The same rule serves another important function: it prevents "feature creep" from prolonging the development and complicating the product. When the FS has been fully implemented, the Design Phase is over. Features that aren't defined in the FS aren't allowed in the product, no matter how attractive they may seem [YOURDON, p. 169, 173].

12.7.3 The FS is the basis for design documents

The first step in the design process is to allocate requirements among the entities that implement those requirements. For certain kinds of equipment, the first iteration may separate functions to be implemented in hardware from functions that software will provide. For other types of products, the first division may be between mechanical and electrical systems. Then these requirements may be further allocated among sub-assemblies, circuit boards, processors, or software modules in a hierarchical fashion. As a concise statement of all requirements, obviously the FS is the essential reference for this process. In the absence of a detailed formal requirements list, no accurate allocation can be done, so the design process is off to a bad start.

12.7.4 The formal documents are the references for design reviews

Designs are typically reviewed at several levels, such as architecture, detailed design, and low-level walkthrough of software code and hardware schematics. The central purpose of all of these formal reviews is to answer the same fundamental question: Does the proposed design meet the requirements of the FS? No matter how attractive a design may be on other grounds, if it fails to meet **any** FS requirement, it must be revised to do so before it can pass the review. To focus on this issue alone until it is favorably resolved, no discussion of functional changes is permitted at the review. When a design is judged to fully provide all mandated functions, then secondary questions can be addressed. Are there ways to improve the design, while still complying with the FS? Can it be simplified, made more robust, manufactured at a lower cost, more easily diagnosed or serviced? All such questions are set aside until FS compliance is ensured.

In informal systems, design reviews (if they are done at all) quickly degenerate into endless discussions about what the product should do. This highlights the lack of earlier functional decisions. An observer at a design review looks for this symptom to assess the degree of formality present. Is the discussion about the implementation of agreed-upon functions, or is it an argument about what those functions should be? In the absence of an FS, there is no standard available to compare against the proposed design, so the principal purpose of the review cannot be fulfilled. The design can be presented, criticized, and admired, but it cannot be measured.

12.7.5 Formal methods produce productivity gains

The SEI Report [SEI2] measures productivity in terms of lines of debugged code produced. A median compounded annual rate of 35 per cent is reported for this statistic. This impressive result clearly reduces the cost of the development and contributes to reduced time to market. Among ISO-registered companies, 69 per cent reported [ISO] improved efficiency.

12.7.6 Formal methods increase early defect detection

The SEI Report [SEI2] investigates the detection and removal of defects during software implementation, before the code is delivered for formal acceptance testing. A median compounded annual rate of improvement of 22 per cent was found. NASA [NASA1, p. 29] also stresses the importance of finding and removing errors as early as possible. Delivering fewer errors to the Test Department clearly reduces the time and effort needed for testing, as well as reducing rework costs when defects are identified after implementation is complete.

12.7.7 The Development Plan is the basis for management tracking

Documents produced early in the Design Phase identify and describe the modules and sub-modules that implement the product. These design documents (AS and HLDS) are the best references for management to use to track the progress of the design. The functional organization of the FS may be used for this purpose until these design documents have been produced and approved. Eighty-six per cent of ISO-registered firms reported [ISO] an improvement in management capability.

12.7.8 Formal methods reduce schedule overruns

Among the benefits of formality is an increase in management's ability to predict the duration of the development tasks. During the Definition Phase, considerable effort is spent to determine the detailed features to be designed. These decisions are recorded in the FS, the basis for establishing the schedule for the Design Phase. The feasibility studies that have been carried out provide insight into possible methods of implementation. Since the scope of the product's features and design techniques are known when the schedule is constructed, the information is at hand to enable a reasonably accurate design schedule to be produced. Significantly less information is available at the start of an informal development, leading to a schedule of lower accuracy.

As discussed in Section 4.8, the preparation done in the Definition Phase reduces the quantity and magnitude of functional changes during the Design Phase. Thus, the probability of timely design completion is increased, and the size of the cumulative schedule overrun is reduced. When a deliberate functional change is made during design, the impact on the schedule is taken into account by the MR reviewers. In an informal system, frequent unanticipated functional changes typically cause the design schedule to be substantially overrun. When a stealth change occurs (see Section 12.1.4), there is no opportunity to adjust the schedule, so the effort needed to perform the change contributes directly to the overrun.

A study [SEI4, p. 7] indicates a reduction in the standard deviation of schedule slips relative to estimated schedule as the maturity of the development process increases. Another study [SEI3, p. 13] found that the fraction of organizations that rated their ability to meet their schedules as "good" or "excellent" doubled when CMM Level 3 was compared with Level 1.

12.7.9 Formal methods reduce development risk

Designers perform feasibility studies during the Definition Phase to determine the feasibility of implementing the functions being considered for inclusion in the product. Presumably, features whose feasibility cannot be established are omitted from the FS. Because of this screening process, the development risk of the remaining features has been reduced. Since the feature set is known in advance of the architectural step, and proposed changes are carefully evaluated thereafter, the probability of producing a kludge is greatly reduced. As discussed in Sections 7.3 and 7.4, the Architecture Specification requires technical risks to be addressed in the High-Level Design Specification. Technical risk may be assessed as an afterthought in informal environments, but there is no mechanism that ensures this important topic will be covered.

12.7.10 The FS promotes specialization of labor

[handwritten: Supports Platform Approach]

The FS serves as an interface between the people who specify the product and those who design it. By completely defining the product's functions, the FS transfers functional knowledge, allowing the defining and the implementing tasks to be done by different groups of people without loss of information. In much the same way, a high-level design document communicates the product's architecture from the senior designers who wrote it to the junior designers who do the detailed development. Each person works at the highest skill level he or she can competently perform. Architects architect, designers design, and authors write. As experience accumulates, individuals are promoted from group to group. The company receives the maximum productivity from its salary dollars, because each employee is always working efficiently.

Another factor is at work here. A larger fraction of the work can be done by junior personnel, because they are provided with "scripts", in the forms of an FS and a high-level design document, to guide their efforts. Thus, the employee pool can be shifted toward more junior and fewer senior workers, leading to a further salary savings.

In contrast, in an informal system senior personnel typically follow a product throughout its development cycle, because there is no other way of providing continuity of information. Much of the time, these senior people are working at tasks far below their skill levels, reducing the firm's productivity.

12.7.11 The FS is used to solicit contract development

A company may want to have its product designed by another firm under a development contract. The contractor's first question is, "WHAT do you want?" The answer is stated in the FS. A high-quality FS enables a contractor to quickly grasp the scope of the design effort, and prepare his bid without padding for contingencies due to lack of understanding of the details required and subsequent surprises.

Since the FS has been carefully prepared and thoroughly reviewed for the benefit of its other users, its cost has already been paid. A company can effortlessly send it with a cover letter to a few design contractors to solicit their bids. If no bid is attractive, the company uses its internal resources to do the design.

From another perspective, the company can easily assess the efficiency of its internal development process by comparing its measured cost with the bids received from contractors to do equivalent work. This competition helps to maintain the efficiency and productivity of the company's staff.

12.7.12 The FS is the basis for a contract

When a satisfactory contractor is found, the FS serves as the basis for the development contract between the parties. (Some aspects of this situation are explored in Sections 8.5.5 and 8.5.6.) Since the FS contains a complete functional description of the desired product, no further functional information should need to be prepared, so work can begin promptly.

If competent contractors are in short supply, the FS may be used as a tool to motivate contractors toward cooperation with the company. Contractors may favor a client with a rigorous FS over another possible client with an informal specification, because the risks of misunderstanding the requirements and unexpected functional changes are lessened by the presence of the FS.

12.7.13 Suspension and resumption of development

At the end of each formal development phase, almost all of the value that the developers have created is captured in the documents that have been released to date, as noted in Section 12.3.4. Thus, the end of a phase is an ideal time to gracefully suspend the development of the product, in such a way that the development could be efficiently resumed at a later time. A new team of developers could quickly get up to speed by reading the formal documents. A powerful motive for planning a deliberate suspension of development projects is discussed at the end of Section 13.3.4.

Resumption of a suspended project is much harder in a informal environment, where the principal repository of information is the minds of the people who started the development. Even if all of the key personnel could be returned to the project, their memories of this product's details have been confused with those of the products they have worked on in the meantime, and some assumptions have been forgotten.

12.7.14 Formal documents assist maintenance

In this sense, "maintenance" refers to the ongoing process of enhancing an existing product after its development is complete; it includes correcting errors, adding capabilities, and tailoring the product for specific niche applications. The FS serves as a baseline, describing the product which is to be modified. During the MR process, the FS is explicitly changed to define the new functionality. These changes inform the developers (or maintainers) to modify the implementation to achieve the new functions, and allow the Test Plan to be changed to verify that the maintenance process had the intended result. The design documentation must be updated in accordance with the implementation changes, so as to continue to represent the product "as built". In this way, both the FS and the design documents are ready to support the next change.

12.8 OTHER BENEFITS

Since every company's situation is unique, some of the benefits listed above may not be available to a particular firm. Since every company's situation is unique, there are probably some additional benefits that haven't been identified in this chapter that can be obtained by a particular firm. Before these benefits can be harvested, they must first be identified. Every developer should be attentive for potential benefits that may be available due to the peculiarities of the developer's formal environment.

13

DECIDING TO UPGRADE TO FORMAL METHODS

Most small and medium-sized companies make use of a wide variety of informal development methods. If these companies could upgrade their practices to operate in a formal manner, quite substantial benefits would be obtained, as described in Chapter 12. If this were an easy transition, it would have already occurred. In fact, upgrading an entrenched informal methodology to a formal one has proven extremely difficult in practice. This chapter addresses the considerations leading up the corporate commitment to make a transition to a formal process. Chapter 14 offers some suggestions for shepherding a firm through this ordeal.

13.1 SHORT-RUN COSTS VS. LONG-TERM BENEFITS

A formal program requires more resources to be invested in the early development tasks than does an informal approach. In the Definition Phase, documents are generated and reviewed, but no tangible design takes place. This is in sharp contrast with the flurry of visible activities that occur at the beginning of an informal development: schematics are drawn, parts are ordered, mechanical drawings take shape, and software coding starts.

This difference is an obstacle to conversion to formality. Management must have the firm conviction that the eventual benefits of the formal approach will outweigh the initial investment, and must not panic because the familiar visible signs of progress aren't initially present. The long-term nature of the benefits must be fully communicated to management, so that immediate profits are not expected. **Figure 12** is helpful for communicating the results that should be expected. It shows the fraction of the design work that has been completed versus time, for informal and formal development methods. In the informal environment, design begins immediately and proceeds until the product is deemed ready for release. In the formal environment, no design is done while the FS is being generated in the Definition Phase. Because formal design is based on this preparation, it proceeds much faster during the Design Phase (shown as the heavy line with steeper slope), resulting in release at an earlier time. This isn't a conjecture based on wishful thinking, but a summary of the actual experiences of real-world organizations developing real products, as reported in Section 12.5.1 above.

13.2 STATED OBJECTIONS TO UPGRADING

Let's examine the reasons that are commonly given to justify a preference for informal methods, to see if any have merit.

13.2.1 Formality takes too long

The automatic response to any suggestion for a change to formal methodology is that formal methods are unacceptably slow. No matter what benefit is claimed for formality, it is "refuted" by reciting the unsupported assertion that informal methods are always much faster than formal ones. Some managers are so unimaginative that this is the only argument they can make against any evolution to formality.

Figure 12. Formal vs. Informal Development

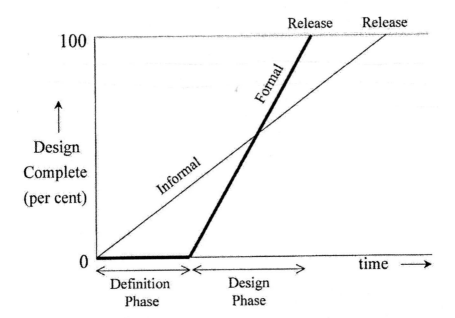

This misinformation is probably based on the assumption shown in Bower's Inequality in Section 3.7 above. The reasoning runs like this:

1. A formal method consist of adding documentation to an informal method.

2. Preparing this documentation diverts resources from design tasks, thereby slowing the development.

3. Documentation plays no role in the informal method, so it isn't used and provides no benefit.

4. Therefore the time spent on documentation isn't recovered, so the formal method takes longer.

As discussed in Sections 3.7 and 4.10.1 above, Step (1) in this sequence is false. If a formal method is applied properly, having robust documents to **drive** the Design Phase accelerates that task so much that the time spent generating the documents is more than recovered. An automatic assertion that formal developments are always slower reveals that the fundamentals of Bower's Inequality are not understood by the speaker.

Design managers frequently want to work informally, or regress to informal methods for the remainder of a project that was started formally, because they are "in a hurry". Each project is seen as a desperate emergency, whose next milestone must be completed as soon as possible. However, the executives who constructed the company's Product Development Procedure knew that this was a universal constant; every step of every project is always "in a hurry". These executives streamlined the Procedures to yield products as quickly as possible, consistent with the desired level of product quality. Therefore, being "in a hurry" isn't a valid reason to try to shortcut the company's Procedures.

The SEI Report [SEI2] shows dramatically that as each organization's methods matured, its overall time to market continued to decrease; a median compounded rate of 19 per cent per year was reported. Thus, there is substantial evidence that formal methods need not be slower than informal ones, and can be faster if properly implemented.

13.2.2 Our situation is special

It is frequently claimed that our company must use informal methods because our situation is so unique that no standard formal method applies. Our product is different, or our customers are uniquely demanding, or we need to keep ahead of competitors, or we must stay even with competitors, or we need to catch up with competitors, or the future survival of our company depends on this development, or the department was just reorganized, or the name of our product begins with the letter "T". We don't need to learn what the formal principles are, because none of them could possibly help us.

One might think that the extreme criticality of a project would be an argument in favor of carefully applying the most powerful, thoroughly tested development method that could be found. This point is never made. Instead, the most amazing *non sequitur* is used to argue that the critical nature of this project compels us to use the weakest possible development method.

Of course, any situation can be presented in so much detail that it appears to be unique. **Because every project is special, no project is special.** Substantial experience teaches that some fundamental processes, such as planning a task before starting to execute it, are universally beneficial. As discussed in Section 11.6.3, the ISO 9001 procedures have been deliberately adjusted to make them applicable to the development of virtually all products and services, based on a mountain of experience with previous generations of slightly more specialized procedures. CMM is also being generalized to strive for wide applicability. An unbiased authority on development methods would almost certainly be able to find one or more standard methods that could be used "as is" or adapted to the "unique" situation at hand to yield significant benefits.

13.2.3 Formality stifles creativity

Several designers have objected to formality on the grounds that it stifles their design creativity. This is a fancy way of saying that they don't want to design the product defined by the FS; they would rather continue to design informally, letting the product's features emerge as side effects of their design decisions. The product is supposed to do whatever the designers say they designed; since the designers establish the product's requirements in this way, it's impossible for the design to be wrong.

In formal systems, an Acceptance Test Procedure driven by the FS is used to objectively measure the correctness of the design. If the product's features differ from those of the FS, the designer's work is regarded as incorrect, and must be changed. This comparison with a reference is an absolutely essential element of a formal program. It ensures that the product supplies the functions that customers are believed to want, rather than functions convenient for designers to implement.

When using a formal method, creativity is encouraged, not stifled. Management demands that the designers (and all other employees) exert their utmost creativity in every task. They must fully exercise their functional creativity during the Definition Phase, and their implementation creativity during the Design Phase.

13.2.4 Formality increases bureaucracy

All formal methodology has been categorically dismissed on the grounds that it increases "bureaucracy". This complaint was made by a bureaucrat in the engineering management hierarchy. He meant that he wanted to continue to work in an unsupervised environment, without any measurement of the quality of his output.

Formal methodology refers to the techniques used for development; it doesn't seek to change the organizational structure of the firm. Complaining about bureaucracy reveals a fundamental misunderstanding of formal methodology. The only additional employees that a typical firm might need to upgrade from informal methods to formality are in the Systems Engineering and Quality Assurance Departments, if those areas have been neglected in the past.

13.2.5 Formality costs more

A summary of the benefits of formality shown by one of the surveys [SEI2] described in Chapter 12 was presented to the engineering management of a medium-sized high tech company. The spectacular advantages in development time, cost, quality, and schedule predictability received no comments. When the cost of a formal program (about one per cent of the total development salary) was mentioned, one VP of Engineering nudged another of the firm's plethora of VPs of Engineering and remarked, "See? There are costs involved." The survey result that showed this cost being recovered many times over through productivity gains was ignored. Clearly, cost wasn't a real issue, just an excuse to delay conversion to formality. The foes of formal methods are grasping at straws in this case.

What is the cost of retaining an informal system while one's competitors are upgrading their methods? If one or more competitors are able to achieve the compounded rates of improvement in product quality and time to market shown in [SEI2], [NASA2], and [ISO], they will gain market share at the expense of laggards. The cost of failing to upgrade may be quite substantial indeed.

13.2.6 What we are using is good enough

Take your best, most recent Functional Specification and honestly evaluate it according to the criteria presented in Chapters 4, 5, and 6. When this exercise was performed recently, a typical FS was found to be deficient in thirteen out of twenty [at that time] categories from Chapters 5 and 6, after giving the document the benefit of the doubt in several others. None of these deficiencies involved spelling or grammar, and each was related to a distinct guideline, so a particular class of flaws was counted only once.

If your evaluation of your FS reveals weaknesses, how can you determine which potential benefits are being lost? In some cases, the relationship between FS characteristics and benefits is straightforward. For example, if the FS contains a manufacturing cost target for the product, the benefits that could have been obtained by disclosing it to persons outside the company are lost. There are many such benefits listed in Chapter 12. In other cases, no direct relationship exists. Each structural defect weakens the FS, increasing the probability of misunderstanding, uncertainty, and confusion. But exactly where is the threshold reached at which the FS becomes unacceptable?

With the list of FS defects in hand, the benefits described in Chapter 12 can be reviewed to see if they are fully provided by the present FS. The total value of the unavailable benefits can be estimated, and compared with the cost of upgrading this FS or producing a more robust FS for the next new product. It is quite likely that the value of additional benefits far exceeds the cost associated with formal methodology.

13.2.7 Small changes require too much effort

Sometimes a small change must be made to a released product, to fix a bug or adjust a feature to meet a customer's nonstandard requirement. It is desirable to make such a minor change in a few days. If the FS, architecture, and design documents must be amended, distributed, and reviewed in turn before a change can be authorized, several weeks may be needed. It has been argued that this situation rules out the use of formal methods for any purpose.

A company's procedures should be designed to optimize the development of major new products, where the bulk of the development funds are allocated and the payoff is greatest, instead of being optimized to facilitate minor fixes. However, the procedures should be flexible enough to accommodate projects of various sizes. The changes to the product must be documented in all cases, using the MR mechanism for changes. If a functional change is made, the FS must be updated accordingly so that it exactly reflects the product being produced. If no functional change is involved (as in a bug fix), the FS isn't changed. Similarly, the Architecture Specification is amended if and only if an architectural change is made. If rapid changes must be made, the procedures should provide for expedited MR processing and product modification before documents are updated. Great care must be taken to prevent these emergency procedures from being misused in non-emergency cases to compromise the integrity of the formal process. A strict time limit (perhaps a few weeks) is recommended between emergency MR approval and document release, to prevent minor changes from accumulating until the documents no longer describe the product.

13.2.8 Our design tool produces adequate documentation

Some computer-based design tools can process the detailed design output (i.e., software source files) to generate volumes of "documentation" that describes the lowest-level design in great detail. This material has been offered as a substitute for all of the documents required by a formal procedure. The designers want to do an informal design, then push a button to transform it into a formal design.

Such after-the-design documents clearly cannot substitute in any way for the documents that are to be generated and **reviewed before** the design begins. The external function of the overall product (the FS) is very different from the sum of the internal functions of the modules that implement the product. Many parts of the Architecture Specification (such as a discussion of alternative approaches and why they weren't used) relate to human decisions that can't be abstracted from the detailed design. Although automatic post-design documentation may be easy to produce, it cannot substitute for the pre-design documents required by the company's Product Development Procedures, for the three compelling reasons just stated. Case closed.

13.3 Actual Obstacles To Upgrading

The stated objections are collectively quite weak, in comparison with the substantial benefits that formality can bring. They should be regarded as superficial reasons for avoiding an upgrade, rather than the actual impediments to improvement, which remain unstated. It's unproductive to try to persuade an executive to approve a process upgrade by demonstrating the weakness of his stated objections. As soon as one "reason" for retaining the *status quo* has been overcome, another artificial objection will be invented to take its place. In order to make progress, the actual unstated obstacles must be addressed. This section identifies a number of possible obstacles and suggests some ways of dealing with them. General techniques are appropriate, because the actual motives of a given group of executives may not be readily discoverable.

13.3.1 Lack of understanding of formal methods

All corporate technical executives and development managers claim to fully understand formal methods. Their average level of actual knowledge is shockingly low, considering that product development methods are the fundamental tool of their profession. How did these executives come to possess the detailed understanding they claim? Human infants aren't born with the knowledge of formal development methods. (In fact, instinctive, untutored behavior leads precisely to the informal approach.) The executives didn't learn formal development in school, because no university offered a course in this subject to the depth presented here. They certainly did not learn formality on the job by watching other executives manage informally; they just learned a lot of bad habits. In the case of most executives, no dedicated study of formal methods has been undertaken. It's as though an aspiring surgeon didn't bother with medical school or the study of anatomy, but just watched some amateur surgeons at work and then picked up a scalpel and started cutting into people.

Every executive has some superficial knowledge of formal methods: some kind of documents are involved and some form of review is needed. As this book stresses throughout, a lot of other interacting elements are necessary to cause a formal program to function properly. Executives who realize there is a body of knowledge they haven't mastered are reluctant to commit to applying formal methods. It's a natural fear of the unknown.

Many companies claim to be using formal methods while actually pursuing "informal-plus" practices. Low-level employees go through the motions of generating and reviewing a thin veneer of documentation (unless their program falls behind schedule), but these documents aren't used to guide later phases. Managers constantly look for "shortcuts", and executives make decisions without regard to the company's procedures. These executives behave as though they believe that complying with 20 per cent of a formal method will produce 80 per cent of the method's benefits. Unfortunately, the reverse is more nearly true. The bulk of the benefits aren't obtained until the paradigm shift (see Section 14.2) takes place, and that occurs after most of the effort has been invested.

There may be cases in which a company actually uses a mixture of informal and formal methods, but its executives truly believe that fully formal methods are being used. This can occur only if the executives have a very limited understanding of formal principles.

Consider a sports analogy. Our team isn't playing very well, and we would like to perform better. When we play against the champion team, we notice that before the contest the champion players line up in a straight line and warm up by stretching in unison, first to the right and then to the left, three times on each side. Then they beat us decisively. After seeing this pattern several times, our team starts warming up in the same way; we line up and stretch in unison, first to the right and then to the left, three times on each side. When a sports reporter asks our coach, "Do you prepare for each game in the same way as the champions?", our coach says, "Yes, we prepare in exactly the same way. We adopted the champion's own preparation method, which we use before all our games." In fact, our team has copied only one visible but superficial aspect of the champion's preparation. We remain oblivious to the important preparatory steps that are responsible for the champion's success: conditioning, practice, study of game films, devising new strategy and plays, etc. Since our team neglects these crucial steps, our overall preparation isn't like theirs at all, so our performance is unlikely to improve. Imitating some superficial aspects of formal development (such as holding mock reviews of low-quality documents) won't lead to any significant benefits.

The obvious **starting place** is to ensure that all those who establish development procedures and implement them have a **sound knowledge of formal methods**, the differences between informal and formal approaches, and the benefits and problems of each. This is a crucial step. Only after a common understanding has been achieved can specific issues be meaningfully addressed.

When corporations want to impart information to their employees, they always use the same technique, which they call "training". An executive or professional trainer prepares slides, then makes one-hour presentations to groups of employees. Training is appropriate and effective for conveying information about low-level mechanical operations: "You unscrew the old bulb counter-clockwise, then screw in the new bulb clockwise. Everybody got that?"

Substantial experience has shown that training is useless when the objectives are to explain formal methods and persuade employees to use them. Managing a technological company's new product development requires integration of a large number of disparate, complex factors; the process is far too complicated to be mastered in a one-hour session. Training might begin to address the stated objections to formality, but cannot attack the actual objections in its public forum. In one example, a knowledgeable, experienced, professional trainer traveled to three continents to train employees of his global firm on changes in their formal development procedure. A few months later, a question was raised: Has any employee taken any action in accordance with the new procedure as a result of the training? No such examples could be found. If no behavior change was motivated by the training, it must be regarded as completely wasted. The costs of the lost person-hours and travel expense are trivial compared with the cost due to the principal problem: the employees still aren't following the new procedure.

Deciding to upgrade to formal methods

We need something much stronger than training. Fortunately, the appropriate tool exists; it's called **education**. The techniques of modern education can be brought to bear: multiple lecture sessions, questions and answers, discussion, thought experiments, case studies, written homework assignments, intensive drill, "laboratory" work, and written examinations. Universities don't say they train their students; they claim to educate them. A university or college is the best venue for a course in formal development methods. The material in this book supports a one-semester course with 45 contact hours. Such a course could also be taught as an intensive, full-time short course of one week, either off-site to participants from several companies, or on-site to employees of a single firm. Top management and all marketing and technical executives and managers must attend, as well as the working marketers, definers, and designers. The ultimate goal is to convince the decision-makers in the company to adopt a fully formal methodology, so they must first be educated about the methods, their implications, costs, and benefits. Applicable surveys and case studies should be used to support the quantitative claims of improvement. Trying to teach the equivalent of a college course in a few hours just leads to misunderstandings and frustration, without persuading anyone of anything.

13.3.2 Disbelief of claimed results

Managers may understand formal development principles and may be aware of the benefits claimed for formality, and yet disbelieve these claims. Informal management is completely intuitive, while formal management often requires counterintuitive decisions. The advantages of formal work are often based on correlations and probable results, rather than on observable cause-and-effect relationships. Throughout the manager's entire informal career, the "slowness" of formal work has been constantly asserted and never refuted, until this dogma has become "knowledge". In the presence of such persistent misinformation, it's reasonable for a manager to question the claims made for formal methods in Chapter 12. Surveys that depend on voluntary responses can't be scientifically rigorous. Companies that have benefited from their transition to formality may be eager to highlight their success, while companies that failed to achieve benefits may be reluctant to respond to the survey, fearing that their poor results may reflect on their competence. Surveys show what is **possible** to achieve, not what is guaranteed. It's always possible to mis-implement (or pretend to implement) the best of procedures in such a way as to produce negative results.

After attaining a full understanding of formal methods, executives should be encouraged to imagine that these methods are fully operational in their firms, and try to visualize the benefits that are gained by this change. A revealing technique is to review a development program that was completed recently (while memory of its details is fresh), to see which of its problems would have been reduced or eliminated if it had been conducted formally. A development project that was discontinued as a failure would be an especially fertile topic for a post-mortem review. If this is done with an open mind, a number of improvements should be apparent. This is analogous to a pathologist performing an autopsy to determine the reason an organism failed, and to a sports team's study of films of previous games to see why they won or lost.

A powerful argument in favor of formal methods comes from the **customers** for the company's products. Thousands of customers, including most of the world's largest technology firms, insist that their vendors must use certified formal development (and production) methods. These aren't case-by-case decisions made by purchasing clerks, but are carefully considered long-term policies established by the firms' top managements, as an integral part of their firms' quality programs. Surely these customers realize that their vendors must pass along the costs of their development programs to their customers, and that a longer development time makes the product available to the customer later. If customers found that higher costs or unacceptable delays resulted from formal development, they could elect to terminate the restriction they have voluntarily placed on their vendors, but they do not. They continue to insist that vendors provide **only** formally developed products, because they know that development techniques continue to influence product quality, even for units manufactured years later. They are so secure in their belief in the superiority of such methods that they refuse to even consider non-compliant vendors. If you have an informal process and a great track record, they don't want to do business with you. If you have a revolutionary new formal technique that works well but is so radical that the recognized quality organizations won't certify it, they don't want to hear from you, either. Prudent people intent on buying the cheapest acceptable item would need to see lots of convincing evidence before taking such an absolute position, wouldn't they?

These knowledgeable customers use formal methods for developing their own products, so they know very well what implications and side effects formality imposes on the developer. Customers demand that their vendors follow through and actually comply with their promised procedures, not just pretend to do so while continuing informal practices. Executives who doubt the benefits of formal development methods should think long and hard about **why** so many prominent customers insist on their use.

13.3.3 Difficulty in getting cooperation

An executive may understand formal principles and believe that substantial benefits will result, yet be reluctant to commit to a formal program because of fears that other departments won't fully cooperate. If others just give lip service to the formal procedures and go through the motions without actually adopting formal methods, confusion and strife among departments will prevent any significant benefits. An executive may also fear that she can't even get her own subordinates to willingly change their operating methods, and may wish to avoid a potential personal conflict.

This is a legitimate concern, since non-performance by a few laggards can nullify the hard work of the rest of the company. For this reason, it's important that **all** marketing and technical executives and top management participate in the formal methodology education program. Peer pressure and the Quality Assurance function serve to detect noncompliance with procedures and steer the offender toward compliance. As a last resort, guidance can be supplied by top management. In order for a quality program to succeed, the Chief Executive/President and Board of Directors must be dedicated, enthusiastic supporters.

13.3.4 Fear of loss of control

Procedures and documents based on formal methods guide many day-to-day decisions: what step should be done next, can a review meeting be skipped, who is authorized to select a subcontractor, which features are in the current release, and so on. In an informal environment, these sources of direction don't exist, so managers spend their time making such decisions on a case-by-case basis. As the transition to formality is made, such micro-management is no longer needed. The apparent lessening of personal control and involvement may be hard on managers' egos.

Formal Quality Procedures are intended to limit the flexibility of executive decisions, steering them along the lines that have been found by other firms to be correlated with success. When executives realize that they, and not just lower-level workers, will be subject to restrictions, they may fear loss of personal control. They will no longer be able to waive or modify parts of the development procedures unless those procedures explicitly give then authority to do so. If they cannot produce creative "solutions" to near-term development problems, their personal importance to the firm appears to be diminished.

Let's consider an example. Suppose a design team unexpectedly becomes available for work before their next product has been defined. This would happen if their previous project were cancelled. [If cancellations are frequent, or if management is taken by surprise when projects are completed, management's fundamental competence is called into question.] In a formal environment, these designers may remain idle for weeks or months while the Definition Phase of their next project is completed. This prospect may motivate an executive to authorize the design team to immediately begin low-level design of the parts of their next product that they believe they understand. Since the FS isn't available to drive this effort, this design must be done informally, forcing the rest of the development into informality as well. The executive in this example is willing to sacrifice all the formal benefits in order to give the appearance that his designers are fully occupied.

When a product is being developed, there are worse things than idle designers. One of them is to develop the wrong product. Another is to lock in inappropriate architectural and high-level design decisions before the functions of the product are known.

Replacing case-by-case human decisions with uniform procedural decisions is a cornerstone of the formal method, and is the source of many of its advantages. This is analogous to the political principle of replacing a "Rule of Man" with a "Rule of Law", which has been shown to be essential for national prosperity and economic growth. This principle cannot be compromised in order to give managers the feeling of personal control.

In formal environments, managers continue to exert control over their departments; they do this **through the company's procedures**, rather than directly as in informal systems. Thus, managers should exercise care as they write and review these procedures. They should try to visualize how the draft procedures will be used in practice, and how they could be misinterpreted or misused, so that these defects can be removed before the procedures take effect. Managers should participate in the company's drive toward continuous improvement of its procedures, to correct weaknesses and reinforce techniques that have worked well.

The use of formal procedures actually **multiplies** the scope of managers' decisions. When an informal manager solves the "problem of the day", she affects one specific problem for one product for one day, because there will be another problem tomorrow. When the same amount of managerial effort is applied to formulating procedures, it may prevent a whole class of problems for all of the company's products for many years into the future. If the scopes of these problems are represented as three-dimensional solids, the informal manager's decision affects a cube one problem high by one product wide by one day deep. The scope of the procedure is many problems high by many products wide by many, many days into the future. Which of these solids has the greater volume, or importance to the company? This is the sense in which a formal manager multiples the value of her contribution to the company by working to establish and improve procedures.

Three scenarios are shown in **Figure 13**. For reference, the top timeline shows the sequence of development tasks mandated by the company's formal procedures; the Definition Phase begins at the start of the project, the Design Phase begins when the FS is available at the completion of the Definition Phase, and culminates with release of the design. The middle timeline shows the situation in the example mentioned above, in which management is surprised by the unexpected availability of design resources. When this occurs, management frantically casts about for a new product concept and seizes upon the first half-baked idea that emerges. In order to employ the designers and finish the development as soon as possible, all idle definers and designers are assigned to the project and told to begin doing their tasks. As discussed in Section 4.10.1 above, this causes the Definition Phase to take longer than it did in the first scenario, and extends the Design Phase even more, so the release of the product is substantially delayed. True, the designers were continuously employed, but much of their labor was wasted.

Figure 13. Three Ways to Schedule Phases

1. Formal

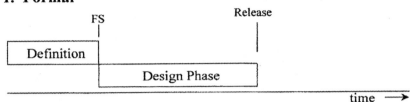

2. Informal plus documents

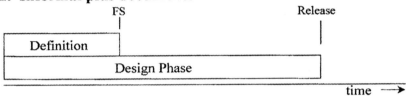

3. Ideal

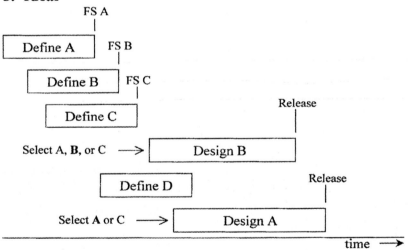

The third scenario shows the opposite case, in which several products are deliberately **defined in advance** of the availability of the resources to design them. Functional Specs are prepared, reviewed, and archived for Products A, B, and C. When designers become available, management makes a selection among the FSs that are ready at that time, choosing Product B in this case. The criteria for this selection may include the urgency of customer needs, projected rate of return, completion of a product line, and maintaining or increasing market share. The formal design of Product B proceeds as in the top scenario. Meanwhile the definition of Product D is started. When another design team becomes available, management selects between the two remaining products whose definitions are complete, electing to begin the design of Product A in this case. Management always **chooses among several attractive products that have been thoroughly studied and defined**; this is obviously a very desirable situation in general, as well as an excellent solution for the case in which design resources suddenly appear. The advantages of formal development are obtained, since the FS is available at the start of the design task, and no employees are idle. The company's procedures should allow for (and encourage) this highly desirable scenario, by not requiring that the Design Phase must begin immediately after the Definition Phase ends. The company should **regulate its ratio of definers to designers** so that a library of alternative products is ready when needed, but that attractive product candidates don't wait in line so long that they become "stale".

13.3.5 Fear of promotion

Formal procedures free marketing and technical executives from making numerous daily decisions. The result is that these executives finally have the time to do senior management activities. They can do strategic planning, communication, research into organizational structures, market studies, administration, recruiting, and (if any time is left) even thinking. The result is that the executives have **promoted themselves** from operating as glorified middle managers to actually doing senior management tasks! To some, this 'promotion' is a wonderful opportunity, but other executives may not welcome this prospect. Most technical executives (i.e., VP of Engineering) were promoted to their present position because of their middle-management skills. They have continued doing the same middle-level tasks after promotion, with a multiproject scope, but with the same orientation. Typically, they have had minimal formal education or experience with high-level executive concepts, such as long-range planning, quantitative financial tools, corporate economics, strategic management, information technology, corporate communications, international marketing, and organizational behavior. They may be reluctant to attempt the actual transition from middle-level technical management to high-level corporate management. They are under pressure to do so because they already have an executive title that they would like to retain. If the other professional managers of the corporation discover that the technical executive has no actual corporate management ability or experience, and that much of his former technical contributions are now handled by routine procedures, they may demote him to his level of competence or replace him.

Competent executives shouldn't fear actual promotion to top management. They aren't expected to run the entire corporation by themselves. Management techniques are learnable skills that an executive can gradually master by self-study or by taking short courses or evening courses, in person or via internet. If the time to pursue an MBA isn't available, the business schools of some universities offer a one-month intensive "mini-MBA" that introduces students to the tools needed to solve many business problems and introduces business and financial terminology so that the technical executive can understand what the other executives are talking about. Technical executives can ask other executives for assistance. "Which textbook is best at explaining financial ratios? May I borrow your copy? What are the implications of P/E=100?"

13.3.6 Fear of commitment and accountability

A formal process requires decisions to be made in the appropriate development phase. These decisions are recorded in the documents that are the deliverables of that phase. The decisions are reviewed when the documents containing them are reviewed. Upon approval, these documents are placed into the document control system. This system distributes the documents within the company, so that they can guide the work done in following phases. A document control system also prevents the author of a document from making unilateral changes to the document later, to make it appear that she was on the "right side" all along. This permanent public record makes an author accountable for her decisions.

Many managers who lack confidence in their ability like to postpone decisions until other workers have made related commitments, so that the manager can appear to be a "team player" by endorsing the consensus, without ever having to go out on a limb by making the first commitment. Their internal monologue may run something like this: "What? You want me to make decisions, write them down, sign my name to them, and distribute copies? Are you kidding? Don't you see how vulnerable I'll be to management's next hind-sighted witch hunt? I've built my whole career by deferring decisions, keeping a low profile, and passing the buck. Now you want me to do exactly the opposite? No way!"

One of the strengths of a formal method is that it forces timely decisions to be made and published, so that later phases work from a common foundation. This fundamental principle cannot be compromised in order to shield managers from criticism.

Top management can do some visible risk-taking of its own, to set a good example and to assure executives and managers that formality isn't being instituted in order to increase their vulnerability. Top management can make it clear that employees will be graded on their degree of compliance with formal methods, not second-guessed when their decisions turn out to be less than optimum. Executives and managers can take consolation from observing that their peers (who they may regard as rivals) are also being compelled to make public decisions and take risks. As benefits are observed, the level of trust among departments should increase, so that defensive behavior is no longer necessary.

13.3.7 Fear of failure

Suppose an executive champions the upgrade to formal methods and the company accepts her recommendation and implements the change. After some time, there may be no observable significant benefits, or performance may have actually decreased. The champion may be blamed for endorsing an inferior method, even though other employees had complied so poorly with the new procedures that improvement was impossible. Faced with this possibility, an executive may lack the confidence to proceed.

In the event that formal development is tried and judged to be a failure, top management will be looking for those executives and managers who didn't fully cooperate or who didn't work hard enough to make the method succeed. The champions of formality shouldn't fear punishment. There are some activities that a manager can do to become a champion (see Section 14.5.4). By taking these actions consistently, a manager will convince his subordinates that they too must follow all procedures. This will seem natural when they do their second formal project; after their third project, they won't tolerate informal methods in future work.

13.3.8 Fear of success

On the other hand, suppose the upgrade succeeds, and the benefits are traceable to the changes in development procedures. Top management might then reasonably inquire as to why the upgrade hadn't been made years earlier. After all, generic formal techniques have been well known for some time, they are in the public domain so they may be used without paying licensing fees or royalties, no reorganization of the firm is involved, and no capital equipment is needed. By not converting to formality earlier, the company's executives implicitly endorsed the previous informal approach, whose deficiencies have just been highlighted by comparison with the new methods.

Executives should divorce themselves from past informal methods by becoming visible champions of the formal program. They should work toward success, as described in Section 14.5, and thereby position themselves to take credit for the benefits of formality. A "better late than never" argument can be made for the current transition to formal techniques.

13.3.9 Admission of past misrepresentation

High-tech companies always claim they use formal development methods. Usually this claim is compelled by large customers who understand the necessity of formality and require their vendors to implement robust development methods. Many companies actually pursue an "informal-plus" strategy; they continue their previous informal development practices while generating a few unwanted documents to support their claim of formality when a superficial audit is conducted. In most cases, the company's executives know that this artifice doesn't actually raise their products' quality, and that they haven't really fulfilled the promises they have made to their customers in this area.

After informal-plus operation has gone on for a few years, an executive cannot recommend a change to formal methods, because the customers, Board of Directors, and stockholders have been repeatedly told that formal methods have been in place throughout that time. A suggestion of a present upgrade is an automatic admission that the upgrade that was supposed to have taken place years ago wasn't actually done. It's much easier to maintain the *status quo*, hoping that the deception won't be discovered.

A transition to fully formal methodology need not be done in a way that involves confessing that the present program has been misrepresented. The change can be described as an "improvement" in methodology, from a formal system to a better formal system. The important thing is to quickly upgrade the development methods that are actually in use, not to quibble over terminology.

Perhaps ammunition is needed to persuade laggards that a truly formal program should be started without delay. They could be asked to document the consequences of the discovery that the nature of the present program has been misrepresented. Discovery could take the form of the failure of a rigorous audit by the outside firm that certifies the company's development methods, or a major customer could ask to see a document that is required by the company's Quality Procedures but wasn't actually written. Such a request could be triggered by the poor quality of the products being delivered, or by a tip from a disgruntled former employee who is now employed by a competitor. If a customer discovers impropriety, the customer can use this information as a bargaining chip to seek concessions from the vendor, such as a much lower price because the products were actually developed informally.

In another scenario, the customer requires vendors to use only formal development methods because the customer's quality procedures mandate this requirement. The customer's procedures may require it to qualify all vendors by auditing prospective vendors, to directly verify that the vendors use a satisfactory method. The only way for the customer's auditor to be sure that the vendor's procedure was actually followed is to review the documents that were produced during development, including the FS and other required architecture and design specifications, minutes of review meetings, test plans and results, and logs from the document control system. Even if the vendor were willing to generate these materials after the development and back-date them to fool the auditor, a great deal of work would be needed, and the probability of producing a consistent document set would be low.

In an OEM situation, the customer integrates the vendor's product as a component within its own design and sells the resulting conglomerate. If the integrator operates under a formal development environment, it must produce a complete, detailed FS for its conglomerate product. Therefore the functions of the vendor's component must be formally specified in detail, to the extent that they influence the external functions of the conglomerate product. Two approaches are possible: (1) the integrator spends a lot of time and effort characterizing the component (with the possibility that the result will be inaccurate, so that the vendor's product doesn't meet the requirement), or (2) the integrator incorporates the vendor's FS into the conglomerate's FS, either directly or by reference. Clearly it's much faster, easier, and safer for the integrator to require the vendor to supply the FS of the vendor's product. If this document isn't available or is of poor quality, the falsity of the vendor's claim that formal development was used is evident. In this connection, it has been remarked that "You can't cook a kosher meal from non-kosher ingredients". The inadvertent use of a non-compliant component has caused the customer to violate its own quality procedures; it may have to report this breach to **its** customers, and **their** customers in turn. The resulting anger may lead to retaliation against the component's vendor.

In summary, if a company pledges to use formal methods but tries to continue informal development with some superficial formal aspects (the "informal-plus" approach), it is in constant danger of exposure from a number of sources. Such misrepresentation could cause the company to lose the certification of its development program, could be regarded as a breach of contract by its customers, and would destroy the reputations of the company and its officers and directors. The prospect of such extreme consequences should motivate the company to promptly take steps to meet its promises and end its liability.

13.3.10 Near-term emphasis

In many small companies, management's attention is focused on the timely completion of the next schedule milestone, and on little else. When judged by this single criterion, short-sighted informal methods that sacrifice everything else in favor of the present task appear superior to formal methods that share their emphasis between short-term and long-range goals. In its extreme form, near-term fixation results in "salami slicing": dividing time into short segments (perhaps a week in length), and announcing that the current task must be completed in the present slice. When it isn't, the pressure is redoubled to finish in the next slice, and so on. The apparent motives for this management style are to prevent loafing and compel unpaid overtime, but its result is to greatly extend the overall development time. Ordinary solutions such as hiring a new employee, buying equipment, and training present employees are ruled out, because they couldn't bear fruit before the "drop dead" milestone at the end of the current time slice. By focusing on a single goal and prohibiting all solutions except overtime, management becomes very easy.

The widespread over-emphasis on achieving the next development milestone by an arbitrarily scheduled date is a serious problem. The success or failure of a product ultimately depends on whether customers place large orders for the product with satisfactory margins. These customers probably evaluate samples of the product first, then introduce small quantities on a trial basis, and finally place the volume orders. These crucial milestones therefore occur many months, if not years, after the development milestone that occupies management's attention. The elements of the development process that affect the customers' decisions are those that **persist** until the time those decisions are made. They include product features, cost, availability, and quality. The factors that influence these persistent elements should be the focus of management's attention during development. At the time of the customers' crucial decisions, the customers literally **do not know or care** whether or not the vendor met some arbitrary internal development schedule milestone months or years earlier; this element isn't persistent.

Formal development focuses on long-term factors, stressing product features and quality. Its success is due in large part to the congruence between its goals and the factors that will be important to customers when they make their volume purchase decisions.

13.3.11 The special case of startups

Short-term emphasis is especially prevalent in startups, for at least two reasons. Before the initial seed capital is exhausted, the development of a product (or service) must be completed, in order to attract secondary financing on favorable terms or to support the company from the cash flow from sales of the product. The second reason is that the startup faces competition from many larger firms with vastly greater purchasing, manufacturing, advertising, sales, distribution, financing, and customer service systems already in place. To succeed, the startup's executives believe they must reach the market first with their new concept, in order to achieve recognition and enjoy a monopoly position until competitors catch up. Those competitors may already be secretly developing equivalent products, to be announced at unknown times, so every day may be important. The startup is in a race against invisible opponents who might be just behind it. In this environment, the temptation to sacrifice large future benefits for small immediate schedule gains is very strong, encouraging informal behavior.

Thus, executives of startups usually feel compelled to neglect documentation and review of their product in favor of obtaining a prototype at the earliest possible date. This decision forces the entire development of the product to be done informally. If some revenue can be derived from this initial product, the next step is to design follow-on products for similar markets by modifying the original product. These descendents of the original product typically occupy the startup's development resources for several years. Since they are based on the undocumented original features, architecture, and design, all of these spinoff products must be developed informally. The only opportunity the company has to make the transition to formality occurs when it finally starts the development of a product "from the ground up"; that is, a product that isn't derived from an existing product.

In one case, a startup threw together a product to address market needs. As years went by, a trickle of these products was sold, as generation after generation of "developers" tried to understand, repair, and improve the original product. Almost all of the firm's "R&D" resources were spent on investigating problems reported by customers, fixing bugs, and adding necessary features that had been overlooked. A negligible fraction of the company's "R&D" resources was available for actual research or product development, which was conducted in an informal manner. This company's product line was surveyed over a two-year period. Some features had migrated from one product to another, but this "high tech" company had developed no significant new features or major products during those two years.

How can such tragedies be prevented? As an absolute minimum, an FS must be prepared for the original product before its design begins. The technical employees must resist the temptation to rush into the lab and start tinkering before the product's functions have been written down, reviewed, and agreed upon. This step isn't a substitute for formal development, but it allows future generations of developers to start with an understanding of the product's goals. The original FS can be updated to accurately define future derivative products as the company evolves toward formal processes.

13.3.12 Avoiding procedure generation

When a corporate decision is made to adopt formal methodology, the first step is for the executives to write the Quality Procedures that will govern future development activities. Since these Procedures will be closely followed, it's important that they are consistent, clearly written, complete, and describe a realistic method that will actually provide the desired benefits. An executive may regard authoring Procedures as unpleasant work, since unfamiliar skills are needed and there is an implicit commitment to eventually comply. It's unlikely that the first draft will be accepted by all the other executives, implying criticism of the executive's draft. There are probably oversights and unintended loopholes in the Procedures that will become painfully apparent when they are put into service. The author may be blamed for these defects, even though they were reviewed and approved by all of the company's executives. Also, the executive may find it difficult to justify her departures from a procedure that she herself had written.

New skills are needed to author formal development procedures, as well as the working documents required for each product by those procedures. These are learnable skills. While an executive develops her skills, assistance may be obtained to expedite the initial work. Authorship of non-critical parts of the documents can be delegated to subordinates, while an experienced consultant may be retained to help with the critical parts. A preliminary Engineering Process Group (see Section 7.1.7) may be appointed and its members may be assigned to help draft procedures. Management should stress that all of the reviewers of a document, not just its authors, bear a shared responsibility for that document's quality. As experience is gained, techniques that lead to robust documents will be retained, while techniques leading to weakness will be replaced.

13.3.13 Inertia

It's always easier to keep doing things as they have been done in the past, instead of changing to a significantly different methodology. It's painful to learn new techniques. A paradigm shift is a new experience; it might be unpleasant. A manager who is comfortable leading informal projects may fear that he won't be good at formal management for some reason. It seems safer to continue to follow the familiar path.

If the prospect of substantial benefits from formality isn't sufficient, it may require some motivation from top management to start the formal ball rolling. Let's keep in mind that there's another aspect to inertia: bodies in motion tend to remain in motion. Once the person, development team, department, or company starts understanding and using formal techniques, the mystery and fear disappear, and the easy course is to continue the transition toward formal methods. The hardest part is starting the transition.

13.3.14 Other obstacles

In general, each manager and executive is aware of some combination of these real or imagined obstacles. They may be present as conscious thoughts or sensed intuitively in a general way. Each company may have its own unique additional barriers to overcome. For example, changing methods may lead to a turf war, or the company may have suffered through a previous unsuccessful corporate re-engineering that was labeled a "formal" process. Although that misadventure was unrelated to the methodology described herein, it left a bad taste in everyone's mouth when they pronounce the word "formal".

13.4 OVERCOMING OBSTACLES

What a long list of formidable obstacles! It's astonishing that any company is able to overcome enough of them to reach maturity. Yet each year many companies somehow manage to make the transition to formal methods.

It should be apparent why the real obstacles aren't publicly stated. The actual motives are all so negative that they can't stand exposure to the light of day; no executive would admit being driven by such considerations. The obstacles can be summarized as: poor knowledge of the fundamental tools of one's profession, mistrust, eagerness to sacrifice corporate goals in exchange for personal convenience and status, concealment of broken promises, avoidance of effort, and avoidance of necessary risk. Clearly, any discussion of these hidden motives would sound like an accusation of wrongdoing, so employees must be very careful when lobbying their supervisors or peers.

However, there are at least two avenues whereby progress can be made. The company's Chief Executive/President and Board of Directors probably don't fully understand formal development principles, but they aren't personally subject to most of the other obstacles. They should be concerned with the long-term success of the company, and would therefore favor formal methods if they knew they were available and weren't being fully used. In order to persuade top management to upgrade to formality, all that needs to be demonstrated is that its benefits outweigh the **stated** objections of executives and managers. The hidden motives cannot be used as counter-arguments, because top management is precisely the group from whom these motives must be most carefully hidden. As noted in Section 13.2 above, the stated objections are largely false and can be overcome with moderate effort. Since customers insist on high product quality and certified development procedures, Marketing is usually an ally of formality, or at least nominal adherence to formal procedures. Once top management is aware of the significant benefits that are available at low cost through fully formal development, as well as the disadvantages of partial compliance, a corporate decision can be made to really, truly institute formal procedures and ensure full compliance. A Quality Assurance Department, reporting to top management, can be formed (if it doesn't already exist) and charged with verifying that the development procedures are adequate and are being followed.

A second avenue to formality lies within the technical and marketing executives and managers who are resisting change. If they read this book or take a course based on it, they will realize that other readers and students are aware of their hidden motives, and that their position is no longer tenable. This occurs within an impersonal relationship between the executive and a book, rather than as a personal confrontation. The employees of every organization have a spectrum of attitudes toward formal methods:

Leader	Prefers	Indifferent	Dislikes	Laggard
Champion of formality	Formality		Formality	Opponent of formality

As executives and managers learn that their actual motives for stalling and opposition are becoming known, they may voluntarily move toward the left side of this spectrum. Or they could realize that if the resources that are being spent to simulate formal development were instead applied toward real formal work, significant benefits would result. Whatever their reason, a voluntary improvement is more likely to succeed than a change compelled by an edict from top management.

14

HOW TO UPGRADE TO FORMAL DEVELOPMENT

14.1 How to Get There From Here

The introduction of a formal development program can be attempted at any time in the development cycle. Two cases are treated below: starting a formal program at the beginning of the cycle (the initial development of the basic product), and converting the company from an informal system to a formal one during the enhancement portion of the cycle.

14.1.1 New product startup

Without question, the best time to introduce a formal development system is at the beginning of the development cycle of the basic product. As the initial product concept is emerging and the team is being brought together, new techniques can be easily introduced, starting with the generation of the FS during the Definition Phase. There are no bad habits to be unlearned and no primordial documentation of poor quality to work around. The FS that is written for the basic product serves as the baseline for the FS for each enhancement.

Management should make every effort to upgrade the company's development methodology when starting a basic product. The basic product and all its derivatives will suffer throughout their lifetimes if inadequate initial documentation is permitted. Another opportunity to start cleanly may not occur for several years.

14.1.2 Existing product enhancement

It is possible to try to upgrade the development methodology during the enhancement portion of the product cycle. The immediate problem is that no competent FS was written for the basic product. Therefore, it's impossible to write a robust incremental FS for each derived product, because there is no baseline to reference. (Other informal documents, such as User's Manuals, test procedures, and sales brochures, may be valuable sources of functional information, but can't substitute for a formal FS. They were written for very different purposes.) Management has to choose between two unpleasant alternatives:

- Delay the development of the derived products while an FS for the basic product and all previous enhancements is generated, reviewed, and approved.

- Try to write an incremental FS to describe the functions of the derived product without describing the detailed functions of the basic product.

The first alternative is distasteful because it postpones the development of the derived product while work is performed that lacks obvious immediate value. This would be the best alternative if it were done early in the cycle, perhaps before any other products had been derived. It would be facilitated if a partial functional description of the basic product were available to serve as the skeleton of the first draft of the FS.

The second alternative leads to a partial FS that violates the Fundamental Rule because it is incomplete. For the rest of its life, the derived product's functions must be defined as "what's in the incremental FS plus whatever the basic product happens to do". The actual situation is far worse than this, because some of the specified features of the derived products may interact in unexpected ways with the undocumented features of the basic product. These fundamental problems prevent a rigorous application of formal methodology, and reduce the benefits that would have resulted had the whole development been done using formal methods. It may not be easy to determine which benefits remain, or the degree of their attenuation. All of these obstacles cause the overall results to be far worse than they could have been had formal methods been applied throughout. The worst possibility is that management may be so disappointed with the results that they discontinue the formal approach, which, of course, had not actually been tried.

14.2 A Paradigm Shift Is Needed

The term "paradigm shift" refers to a revolutionary change in the model that is used to describe the fundamental nature of something. Well-known examples are the change from Newtonian physics to relativistic physics, and the change from Newtonian physics to quantum mechanics. In both cases, new basic axioms are introduced and numerous consequences are derived. Making only a portion of these changes isn't productive; there is no gradual approach to the transition. The new concepts **replace** a portion of the previous model.

Although "paradigm shift" has been over-used to refer to ordinary changes, it fully applies to the change from informal to formal development. A fundamentally different approach to development is required, affecting all departments.

To recapitulate: in informal environments, there is only one development phase. Armed with a general idea of the desired product, the designers make architectural and design decisions that determine the features of the product. During this process, marketers contributes a sequence of requests for new features, causing numerous redesigns. The product is delivered with whatever features are believed to be working when the emergency extension to all of the normal extensions to the development schedule has been exceeded.

Formal development divides the process into two parts. In the first part, a detailed definition of the product is synthesized. **Then** the defined product is designed to precisely follow this functional definition, with an absolute minimum of feature changes authorized during design.

The formal method is fundamentally different in two ways. First, the Design Phase isn't started until the features are finalized. Second, the design is limited to the specified features. These both sound like simple changes, but both require enormous discipline to achieve when one is accustomed to operating informally. It's very difficult to incrementally modify an informal method to yield a formal method. The informal concepts that were followed without being aware of their existence must be discarded and replaced with new principles. This paradigm shift extends to Marketing and Engineering, and from top management to the lowest level workers. Everyone must change his or her activities. Partial compliance is ineffective; the entire paradigm must shift to secure full benefits. This is never an easy transition. [KOTTER] is an excellent general reference to guide such a process.

14.3 Establishing Development Procedures

After a corporate commitment to formality has been made, the first step is to formulate the Quality Procedures that will govern the development processes to be used. These procedures are influenced by the characteristics of the industry, company, corporate organization, personnel, products, and government regulations that apply to this specific situation. For this reason, detailed procedures can't be stated herein. However, a number of principles are given in Section 7.1 to guide the formulation of the company's formal Product Development Procedure. It may seem at first that too many details are required to be rigorously spelled out; these provisions are needed to prevent the formal principles from being neglected or subverted. Compromise on any essential principle will cripple the entire program and greatly attenuate its benefits.

14.4 START WITH A ROBUST FS

There is one crucial point that must be made concerning the sequence of events when starting a formal program, or converting an informal environment to a formal one. The essential first step is to develop a competent FS. To stress this issue, let's draw a caricature of a small company that starts its conversion process in exactly the wrong way.

Let's Skip the Definition Phase

The company president delivers his kickoff speech to his workers. "Here's the specification for our next new product. It's a two-page list of the product's main features in bullet form, like a sales brochure. See how detailed it is? For example, it says

- For residential use

   so that defines its environment. Then it says

- Meets all applicable Government safety standards

   Safety's good, right? I wrote

Functional Specification

in big letters at the top, signed it, and posted it on the bulletin board in the lunch room. Go ahead and develop a product based on this FS. I want to do this right, so we can't have any sneak paths; no one is allowed to ask anyone any functional questions. If you don't know what the product is supposed to do, go to the bulletin board and read the FS. It seems adequate to me. Shucks, my brother and I used to develop and sell products based on a single sheet of paper. Let's begin our new formal program today, and let those dozens of benefits listed in Chapter 12 start rolling in!"

We can confidently predict one benefit: this company will go out of business soon, releasing its resources for use by sane managers of other firms. The bulletized feature list is laughably inadequate to direct the company's activities; it probably doesn't meet **any** of the criteria listed in Chapter 5 for a competent FS.

The point is that you must start with a detailed, robust, thoroughly-reviewed FS. You can't base a large fraction of the company's operations on an incompetent FS, and hope to iterate toward success. You won't be in business long enough for this to occur. Chapters 4, 5, and 6 describe in detail how to generate a suitable FS. Review techniques specifically for the FS are in Section 8.5.

14.5 Management's Role

A formal program must be instituted by management in a top-down fashion. It must have the enthusiastic, dedicated cooperation of all executives. No amount of careful compliance by subordinates can overcome management indifference. A study [SEI3, p. 23] reported that management commitment to process improvement is highly correlated with success.

14.5.1 Understand formal principles and their implications

Before management can commit the company to a formal development program, the managers must understand that such a program has widespread and long-lasting effects; it isn't just doing more careful documentation during development. Executives and managers must be willing to fully support the formal program by their personal actions, as described in the sections that follow.

14.5.2 Evaluate costs and benefits

It is important for management to understand the nature of the proposed changes to the company's operations before estimating the costs and rewards of adopting a formal methodology. As discussed in Section 4.10.1, a formal program doesn't necessarily require employees to perform more work than they have been doing in the informal environment. It requires them to do different work, which may take more time or less time than before. What is certain is that the early stages of the formal development will appear to move slowly, since the design is postponed until the requirements are finalized. If the formal program is performed properly, this time will be more than repaid over the life of the product, in the form of less rework, fewer bugs to correct, and less repair of products that have been delivered to customers. The principal costs of a formal system are likely to be the time spent training employees to understand and use formal principles, and in the creation and operation of the document control system.

The general benefits to be expected from the formal methodology are presented in Chapter 12. A particular company's product, industry, or special circumstances may nullify certain benefits, but other benefits may be found to replace them. Many of the benefits are intangible, but their dollar value can be estimated after thoughtful consideration of the likelihood that beneficial opportunities will arise and the value of being prepared to exploit them.

14.5.3 Decide and commit to that decision

Once a corporate decision to start a formal development program has been made, the executives must signal their commitment to the new program by every word, thought, and deed. This is crucial as the program starts, because employees always test new corporate policies to see if they are really expected to comply. Is this like the rule that was posted last year, solemnly forbidding all employees from parking in the north parking lot? On the next day, the employees watched carefully to see where the executives parked. The executives continued to park in the forbidden lot, and let it be known that the rule was required by the insurance company's lawyer. No one has paid the slightest attention to the rule since, but it is still technically in effect.

Clearly, partial compliance to a formal program is ruinous. Just one sneak path weakens the whole program, and several such leaks reduce it to an informal program. If an executive makes "harmless" jokes about the new program ("We'll be up to our hips in documents soon."), he is parking in the north lot. If any executive gives the impression that the program is for low-level workers to follow but doesn't apply to the executive herself, the program is doomed.

14.5.4 Endorse, teach, motivate, and enforce formal procedures

Employee confidence is fragile. For example, after employees "buy in" to the concept that the FS will be stable during development and base their own work on that premise, they will rightfully be upset if changes to the FS cause them to re-do some of their previous work. A very few such episodes will convince them that the new methods don't work, or that management lacks the determination to carry through on its promises.

Therefore, it is essential for management at all levels to make a concerted effort, every day and in every way, to cause the underlying premises of the formal program to be met. Foremost, management must set a good, no, a **great** example by the personal activity of the executives. As the new system is being learned, constant vigilance is needed to prevent departures from the plan, and promptly correct the transgressions that are detected. This sort of thing is called Leadership.

The goal is to obtain enthusiastic cooperation with formal methods by employees throughout the company. This is a challenging task, because the company (or the employee's previous firm) has plagued its workers with a parade of "revolutionary" programs that required them to modify their behavior in various ways. These faddish programs may have been called Management By Objective, Quality Circles, or Decision-Making by Scissors-Paper-Stone. The employees learned to use passive resistance to cope with each new program. They gave lip service to the program's rituals and went through the required motions, while trying to perform their duties by informal methods on the side. Without true cooperation from the employees, each new program collapsed and was replaced by another.

We need to do **much** better than this. The first step is to teach the methods of the formal development program carefully, motivating each departure from the informal approach. The benefits listed in Chapter 12 can be explained, and the way each benefit results from adherence to the formal program can be illustrated. Employees should be taught the basis of the formal plan, as presented herein, and its operation throughout the company, as well as the specific rules governing their individual jobs. The overview allows them to understand the integration of activity that a formal program involves. They can appreciate that a departure from the procedures that seems harmless from their local perspective may be harmful to employees in other departments.

Management has several tools at its disposal to facilitate the learning process. The Product Development Procedures should stress the **uses** of the various documents that it requires, not just mandate their existence as an end in itself. Training sessions should emphasize the principles set forth in Sections 3.7 and 3.8, and the multitude of benefits that formal methodology provides. In particular, the formal documents aren't being written just for the designers, but will be used by people throughout the company for many years, providing the benefits listed in Chapter 12. Management should make it very clear that it doesn't expect "extra work" to be done within the same schedule. Management does expect the developers to work **differently**, replacing their present methods with formal techniques. Management should reiterate its understanding that the formal system involves more up-front investment of effort, and pledge to provide adequate schedule and budget. A longer development schedule may well be appropriate as the transition to a formal program is made, since the efficiency of the new methods may be lower than normal as they are being learned.

Managers can help formulate and review the company's Quality Procedures. After the procedures have been put into effect, managers must convince their subordinates that the company is serious this time; all procedures must be followed exactly in all cases. To make this point, a manager can participate in the education process whereby all employees are taught to use the new procedures. He can conduct supplementary seminars for his subordinates. The single most important step a manager can take is to scrupulously follow the procedures himself. The manager can point out cases in which he takes the path required by the procedure when it differs from the path that he would have taken under the old informal regime. (Of course, this must be done positively, not as "Look! I'm being forced to make this mistake.") A manager must use the terminology and concepts of the procedure in all verbal and written communication. To ensure that communications stress the formal concepts, each manager can establish a personal policy of using formal terms and principles at least five times each business day, when communicating with a subordinate [KOTTER, p. 94]. A manager should prepare for, attend, and participate in the review meetings required by the formal procedure. A manager can identify (and quantify, if possible) the benefits from formality as they occur and publicize this information.

Management should point out that designers are automatically rewarded for learning formal techniques. If the formal program meets its essential goal of reducing the number and scope of changes to the product, the designers will spend more of their time working on exciting original development and less on the drudgery of reworking and re-

reworking old products. As the designers learn new skills of writing and reviewing formal documents, they will become more productive, hence more valuable to their employers. Management should reward those designers who successfully make the transition to formality. The designers will soon realize they have made themselves much more attractive to other employers who use formal methods, or who would like to start doing so.

When a manager interviews candidates for employment, he can ask them about their prior experience with formal development, and selectively hire those who have prospered in formal environments. We certainly want to exclude laggards who will impede our progress toward formality. Related interviewing considerations are discussed in Section 12.2.3.

When performance reviews are done, a manager must emphasize that employees who have complied with the new procedures are rewarded and promoted, while those who have not are disciplined and demoted. This lets employees know that management is serious. If there is no feedback for good or poor compliance with procedures, why should an employee bother? In order to distinguish actual compliance from pretense, a manager must monitor performance carefully throughout the evaluation period, in full cooperation with Quality Assurance. A manager can observe his peers' degree of compliance with procedures, and complain to Quality Assurance or top management if noncompliance negatively impacts the activities of the manager's department.

14.5.5 Keeping score

Managers need to monitor the formal program in two ways:

- To what degree are employees complying with the Procedures?
- Are the expected results being obtained?

Clearly, compliance must be attained before management can expect the "expected" results of that compliance. As employees use their formal quality program, many of their decisions are guided by the principles and Procedures of the program. In many cases, the result of a decision is the same as the old informal system would have produced. In other cases, the formal program will yield different results, generally requiring work to be done sooner or decisions to be made earlier. When the formal system generates the same result as the informal system, it hasn't contributed any value; it has just told you to do what you would have done anyway. The only important cases are those in which the results of the two systems differ. The formal method delivers its value in these cases, by steering employees away from destructive intuitive practices toward techniques that have proven valuable to many others under similar circumstances.

There's a strong temptation to continue using informal methods while pretending that formal methods are being followed. Suppose an executive prominently announces that he is using formal methods whenever the two methods call for the same action, but quietly takes the action directed by the informal method when the two methods call for differing actions. This is clearly identical to following the informal method all of the time, while totally ignoring the formal method. However, such a program of selective publicity could cause a naïve observer to believe that formal methods were always being used. Perhaps the executive believes that the "credit" he has accumulated by following the formal path in trivial cases can be "spent" to excuse his noncompliance in important cases.

To measure actual compliance, management should be alert for **all** departures from the approved Quality Procedures, and count their occurrences. Are the required documents being generated? Are their contents adequate? Are they ready at the proper stage in the development process? Are required formal review meetings being held? Do reviewers prepare adequately beforehand? Do constructive, productive, meaningful reviews take place at these meetings? If a review fails, is it repeated after its subject material is upgraded? Are design tasks beginning before the FS is approved? Are formal concepts and terms being used in conversations? Is the document control system working properly? Who approved an exception from the Procedures? Is that person authorized by the Procedures to take that action? If management needs help in monitoring compliance with procedures, an agency that certifies quality programs can be hired to conduct an informal "pre-audit" before applying for formal certification.

How can we determine whether the necessary paradigm shift has occurred? Before the shift, managers conduct their tasks without reference to the concepts and terminology of the formal procedures, and lower-level employees grudgingly follow the letter of these procedures. After the shift, managers and employees willingly follow the letter and spirit of the procedures, constantly using the formal concepts and terminology.

Once substantial compliance has been attained, management attention turns toward measuring the results of the formal program. Quantitative measurements of activities and results are the focus of CMM Level 4 (see Section 11.2.4). One of the key near-term results is that the quantity and scope of functional changes during development should be reduced, as contrasted with the informal system. The MR system described in Section 4.8.4 highlights these changes. The corresponding changes that occurred under the informal environment were not similarly documented, and may not have been recognized at all (particularly if you are working with Ralph or Susan). Therefore, a thoughtful evaluation of the informal changes, including feature creep during design, must be made before a fair comparison between the two systems is possible.

The number of net functional changes can be found in an informal environment by comparing the first draft of the product's spec with the final feature set provided by the released product. But during the design process, a particular feature may have started as Version A, changed to Version B, returned to Version A, then changed to Version C, and then changed to Version D. When the original version (A) is compared to the final version (D), one feature change is counted, whereas there were actually four such changes during the design process.

In addition to tabulating the raw number of changes, the reasons for functional changes should be investigated. If an unforeseeable change in the marketplace forced a change to a product, this is an unfortunate but unavoidable consequence of a dynamic marketplace, so no corrective action need be taken. If frequent changes are the result of inattention during the Definition Phase, appropriate feedback must be applied to ensure that this crucial phase receives concentrated effort from all parties during the next new product's development.

Changes in other characteristics of the development process can also be observed and tabulated. Improvements in time to market, development cost, and schedule predictability can be measured. Product quality can be inferred from the quantity of bugs that are found by customers. Measuring the performance of the development process plays a large role in all formal methods, and should be started as soon as possible in the transition to formality.

14.6 Summary

There is a distinct difference between informal and formal methods. An experienced observer can readily determine the approximate maturity of a given development group. Formal methods offer significant advantages that have motivated a growing number of small and medium-sized organizations to upgrade their methodology. Regrettably, many other such organizations continue to resist making the transition to a fully formal development environment. A substantial list of stated and unacknowledged objections must be overcome before an informal organization will commit to improving its methods.

BIBLIOGRAPHY

[DUNN]   Dunn, Robert H., *Software Quality: Concepts and Plans* (Englewood Cliffs, N. J.; Prentice-Hall, 1990).

[ISO]   Weber, Cynthia, "Benefits of ISO 9000", slides, Quest Analytical, Inc., 1999.

[KOTTER]   Kotter, John P., *Leading Change* (Boston, Mass.; Harvard Business School Press, 1996).

[NASA1]   NASA Software Engineering Laboratory, "Recommended Approach to Software Development", SEL-81-305, Rev. 3, 1992.

[NASA2]   Software Engineering Institute, "Software Process Improvement in the NASA Software Engineering Laboratory", Technical Report CMU/SEI-94-TR-22, Carnegie Mellon University, 1994.

[NASA3]   NASA Software Engineering Laboratory, "Software Process Improvement Guidebook", SEL-95-102, Rev. 1, 1996.

[SEI1]   Software Engineering Institute, "Key Practices of the Capability Maturity Model", Version 1.1, Technical Report CMU/SEI-93-TR-025, Carnegie Mellon University, 1993.

[SEI2]   Software Engineering Institute, "Benefits of CMM-Based Software Process Improvement: Initial Results", Technical Report CMU/SEI-94-TR-013, Carnegie Mellon University, 1994.

[SEI3]   Software Engineering Institute, "After the Appraisal: A Systematic Survey of Process Improvement, its Benefits, and Factors that Influence Success", Technical Report CMU/SEI-95-TR-009, Carnegie Mellon University, 1995.

[SEI4]   Software Engineering Institute, "The Capability Maturity Model for Software Version 2B", slides, Carnegie Mellon University, 1997.

[SEI5]   Software Engineering Institute, "Process Maturity Profile of the Software Community 2001 Mid-Year Update", Carnegie Mellon University, 2001.

[YOURDON]   Yourdon, Edward, *Structured Walkthroughs*, Fourth Ed., (Englewood Cliffs, N. J.; Prentice-Hall, 1989).

Index

Numerics
33-Day Miracle, 10, 15

A
Acceptance Test Phase, 117, 122
Acceptance Test Plan, 52, 57, 69, 116, 156, 157
Acceptance Test Procedure (ATP), 61, 79, 80, 97, 110, 177
addition of documents, 18, 56, 175
algorithm, 94
allocation, 92, 113, 129, 165
analysis, 9
applicability, 136
approval, 8, 13, 17, 30, 32, 39, 47, 49, 51, 82, 88, 90, 94, 98, 103, 138, 147, 180, 202
architecture, 9, 56, 169
Architecture Specification (AS), 50, 85, 89, 91, 92, 94, 95, 96, 98, 111, 115, 167, 168, 180
Architecture/Planning Phase, 115
authority, 85
authors, 28, 31, 39, 64, 66, 67, 70, 76, 78, 80, 83, 87, 88, 96, 98, 100, 103, 105, 147, 169, 194, 201, 202

B
benefits, 57, 140, 142, 148, 151, 155, 157, 163, 164, 173, 186, 212
black box, 19, 65, 98
block diagram, 65, 91, 96
blueprints, 14
boundary definition, 65
Bower's Equation, 18
Bower's Inequality, 17, 56, 87, 175
bullets, 78
burn-in, 104
business models, 62

C
Capability Maturity Model (CMM), 128, 137, 138, 141, 177, 218
certification, 123, 156, 198
champion, 4, 195, 204
change, 7, 14, 19, 33, 34, 54, 55, 76, 146, 147, 160, 171
check and balance, 53
college course, 6, 182, 185, 193, 204
color, 106
Compliance Matrix, 112, 155
consistency, 70, 95, 96, 112
construction industry, 14
consultant, 31, 58, 104, 202
Contract model, 62
contractor, 74, 108, 110, 170
contradictions, 79
cooperation, 54
correlations, 11, 12, 17, 185
counterintuitive, 16, 185
creativity, 55, 177
customer, 3, 7, 14, 21, 22, 28, 31, 33, 36, 39, 50, 51, 55, 57, 58, 60, 62, 63, 65, 66, 67, 69, 77, 81, 104, 105, 111, 123, 125, 126, 135, 147, 156, 159, 160, 161, 186, 187, 196, 197, 199
Customer Service, 29, 163

D
decisions, 29, 75, 82, 97, 103, 121, 189, 194, 212
Definition Phase, 3, 12, 14, 26, 29, 31, 45, 47, 49, 55, 58, 59, 63, 71, 82, 114, 115,

118, 143, 150, 157, 167, 168, 173, 178, 189, 190, 206, 210, 219
delay, 57
design modifications, 9, 14, 35, 36, 50, 56, 89, 93, 143, 161, 168, 171, 180
Design Phase, 3, 12, 14, 37, 55, 60, 63, 74, 75, 91, 95, 96, 97, 114, 118, 136, 147, 162, 165, 167, 174, 176, 178, 190, 209
designers, 29, 55, 118, 164, 169
Detailed Design Phase, 116
Detailed Design Specification (DDS), 85, 89, 96, 111
details, 66, 95
Development Plan, 75, 89, 97, 115, 167
document control system, 37, 90
documents, 13, 18, 37, 56, 64, 68, 71, 83, 86, 87, 88, 89, 97, 102, 111, 128, 140, 155, 175, 180, 183, 197, 213
done and tested, 56, 57
drive, 2, 13, 17, 18, 55, 63, 96, 98, 101, 115, 122, 125, 176, 177, 189
due diligence, 48
duplication, 76

E

education, 185, 214
Engineering Department, 164, 209
Engineering Process Group, 91, 202
executives, 4, 32, 40, 48, 49, 54, 57, 82, 85, 121, 152, 176, 181, 182, 185, 187, 188, 189, 192, 195, 200, 202, 204, 211, 213, 217
expansion, 50, 93
extra work, 56, 214

F

feasibility, 29, 37, 118, 167, 168
feature creep, 51
feature-driven, 15
feedback, 31, 49, 216, 219
flashing indicator, 34
flexibility, 84, 93, 180, 188
formal, 3, 11, 12, 13, 18, 87, 126, 138, 141, 173, 174, 206, 209, 211
formal language, 67, 76, 107

Functional Specification (FS), 13, 14, 15, 19, 25, 28, 30, 39, 47, 55, 62, 64, 73, 88, 104, 115, 138, 147, 179, 180, 210
Fundamental Rule, 25, 40, 41, 45, 48, 60, 64, 164, 207
Future Feature List, 51, 75, 99
future tense, 67

G

government standards, 135

H

hand washing, 12
help-wanted ads, 150
High-Level Design Phase, 115
High-Level Design Specification (HLDS), 85, 89, 95, 96, 111, 167, 168
HOW, 19, 74, 85, 159
Hybrid model, 63

I

IEEE, 134
Implementation Phase, 116
in a hurry, 17, 35, 176
incremental FS, 207
informal, 3, 7, 54, 56, 138, 150, 166, 174, 208
informal-plus, 3, 14, 18, 56, 128, 182, 196, 198
ingredients, 138
interfaces, 13, 19, 30, 57, 61, 65, 66, 68, 91, 93, 95, 96, 169
Internal model, 62
investors, 152
ISO 9001, 126, 136, 141, 177

J

Junkyard Wars, 10

K

Key Process Areas (KPA), 128, 137
kludge, 9, 33

L

lack of understanding, 17, 78, 84, 170, 178, 182, 183
laggard, 4, 153, 179, 188, 197, 204, 216
Levels (CMM), 128, 137, 148, 168
levels of abstraction, 35
library, 24

M

management, 5, 8, 10, 54, 58, 147, 174, 193, 211, 213, 216, 218
management focus, 8, 15, 198, 199
marketers, 7, 28, 55, 69, 99, 118, 185, 208
marketing, 27, 28, 30, 31, 45, 54, 55, 71, 91, 147, 157, 159, 163, 185, 188, 204, 209
minutes, 99, 103
Modification Request (MR), 37, 39, 44, 47, 48, 51, 98, 147, 180, 218
motivation, 57
MR reviewers, 37, 39, 40, 44, 47, 48, 99, 147

N

NASA, 133, 141
no-brainer, 16
nomenclature, 88, 92, 95, 96

O

open issues, 82, 97
outcomes, 103
over-specification, 80

P

paradigm shift, 183, 208
partitioning, 91, 95
persistent, 199
persuasion, 94, 96
phases, 87, 114
post-mortem review, 186
product development, 3
Product Development Procedure (PDP), 47, 56, 59, 83, 85, 86, 87, 90, 91, 101, 103, 115, 122, 147, 176, 210, 214

product's name, 71
productivity, 166
promotion, 192
proposal, 77, 161
proprietary information, 82
prototype, 8, 59
pseudocode, 96

Q

quality, 1, 15, 45, 47, 58, 101, 111, 120, 128, 148, 155, 156, 176, 179, 186, 188, 196, 199, 219
Quality Assurance, 91, 121, 130, 178, 188, 204, 216
Quality Procedures, 83, 86, 120, 149, 151, 188, 197, 201, 209, 215, 217

R

Ralph, 144
recipe, 138
reference, 13, 17, 57, 68, 70, 91, 92, 95, 96, 98, 102, 155, 162, 165, 178, 207
research, 21, 69
reuse, 92
review, 17, 28, 49, 83, 87, 90, 92, 94, 96, 100, 104, 111, 115, 116, 121, 131, 136, 165, 180, 181
review meeting, 32, 89, 97, 99, 100, 101, 111, 147, 215, 217
reviewers, 31, 39, 90, 147, 156, 162
risk, 94, 95, 168, 194
rogue developer, 51
role playing, 104
Rule of Law, 189

S

salami slicing, 199
schedule, 7, 8, 15, 56, 57, 82, 97, 167, 199, 215
schedule-driven, 15, 198
section numbers, 43, 69, 71, 92, 95, 98, 112, 160
selective emphasis, 78
self-study, 6

sequence, 13, 87, 114, 137
shall, 67, 70
short course, 6
sneak path, 41, 159, 213
sports analogy, 183
standards, 68
startups, 124, 200
stealth change, 146, 151
sub-sections, 68, 69
survey, 141
Susan, 44
systems engineers, 5, 28, 31, 69, 99, 118, 137, 178

T

Test Department, 29, 167
test items, 76, 79, 80, 97, 98, 110, 117, 157
Test Report, 98, 118
textbook, 6
time to market, 122, 158, 166, 176, 179, 219
top management, 4, 39, 85, 151, 185, 186, 188, 193, 194, 195, 202, 204, 209, 216
top-down, 14, 70, 128, 211
traceability, 13, 92, 95, 113, 155
training, 149, 184
transition, 4, 6, 17, 28, 39, 54, 55, 56, 153, 173, 186, 188, 196, 200, 203, 208, 209, 215, 216, 219

U

user, 66, 105

V

validation, 122, 127
verification, 122, 127
visibility, 144
visualizing the product, 105

W

WHAT, 19, 56, 66, 73, 85, 159, 170
Wish List, 28, 45, 91
work differently, 212, 215

0-595-27185-5

Formal Development   Wish List
                         ↓

Definition Phase   RS
                         ↓

Design Phase ⎤   AS   Deploy   ~~ATP~~
              ⎥   ↓

    PDP ⎥   HLDS
             ⎥   ↓

        ↓   DDS
             ⎥   ↓

             ⎥   Imp
             ⎥ |

        ↓   ATP — For Every Shall There Is A Test ~~Hen~~

Notion Of Compliance Matrix Pg 112